Yogesh,

My dream for this book is that you will read it and resonate with a few of the ideas. You and I grew up in a time before technology so we are uniquely suited to help the next generations learn to maximize the positives + minimize the negatives. I so much appreciate your friendship the past years and look forward to many more!

Scott Crosby

2016

DID GOD CREATE
THE INTERNET?

The Impact of Technology on Humanity

SCOTT KLOSOSKY

Did God Create the Internet?

CrowdScribed
3540 S. Boulevard Suite 100, Edmond, OK 73013

Published in the United States by CrowdScribed, LLC

Copyright © Scott Klososky

Library of Congress Cataloging-in-Publication Data
First published 2016.

Typeset by CrowdScribed in conjunction with Lightning Source, La Vergne, Tennessee. Printed in the United States by Lightning Source on acid-free paper.
Set in Adobe Garamond Pro

Humalogy™

ISBN 9780996964975
Ebook available.

Technology is a gift of God. After the gift of life, it is perhaps the greatest of God's gifts. It is the mother of civilizations, of arts and of sciences.

Freeman Dyson

DEDICATION

This is more than just a book to me. This is a long discussion I have wanted to have with you (and the world) for many years. This represents thousands of hours of work by many people, and without their help and support you would not be reading these words. I thank my wife, Annette, and our kids for being supportive and always answering my questions when I needed their opinions. They are the ones who pay the price of those thousands of hours when I am head-down in a laptop. Thank you to my mom, Nancy Glende, for hours of editing and conversation that helped me weave together the digital and spiritual worlds, and, for making me, take commas out, of so many, sentences. Thank you to the team at Future Point of View for your ideas and input on this book, I could not do what I love without all of you helping me. The original manuscript was much larger than what you have here, and I have Bill Crawford to thank for being willing to cut it down to the important ideas. And finally, a big thank you to Christian Piatt for doing all of the final proofreading and providing last minute improvements and Brianna Spayd for helping with the mechanics that turn a Word document into a finished book.

I am a very small part of a larger system…

TABLE OF CONTENTS

THE DIGITAL
AND THE
SPIRITUAL

"Sometimes I think we're alone. Sometimes I think we're not. In either case, the thought is quite staggering"

—Richard Buckminster Fuller

In 2015 Pope Francis stated that texts, Internet, and social media are a gift of God. He went on to say they are "a gift of God which involves a great responsibility." I found his comments timely for the topics in this book!

Both the blessing and danger with technology derive from it being a more powerful tool than anything invented up to this point. Never have we had an ability to press a button and instantly connect with billions of people anywhere in the world. Never before have we had the ability to archive the majority of human knowledge in one place while also offering the means to access it instantly. We are putting this life changing digital toolset in the hands of billions of people and hoping that nothing but progress will come of it. Unfortunately, our past does not suggest that only positive outcomes are in store for us as we immerse in a digital world.

While on this planet, we journey through our lives and, at least some of us, seek progress through either the pursuit of wisdom, the pursuit of enlightenment or both. Others merely exist and survive without giving much thought as to why they are brought here or what they are meant to do to improve the world. The remaining small percentage have bad intentions and no problem destroying humans and property for whatever reasons drive them.

Across all aspects of humanity there is the possibility of creating a spiritual connection that can guide how we operate. I believe in a higher being who gave (gives) us all life and that we have a soul that seeks to grow and be connected to others. To the extent that we mature in the connection with our soul and the higher power, we become more enlightened and mature as human beings. As we mature, our soul can awaken to the Truth (the absolute truth, not a perception of the truth) about our world and our lives. In finding the Truth we take a major step forward for humanity.

The question posed in the title of this book, "Did God Create the Internet?" is a way for us to address the role technology has in helping us achieve a more enlightened human existence. One unavoidable reality of this question is that when we invent powerful tools, how we choose to use them (for healthy or unhealthy activities) has massive implications. Ultimately their integration into our lives could be the trigger for the next level of spiritual awakening for humanity or it could slowly leach away what is human in us to such a degree as to cripple our lives.

There are many interesting aspects at the intersection of the digital and spiritual. For example, we have built the backbone of the Internet with the installation of fiber optic cables. Therefore, the transmission vehicle for much of our online capabilities is light. In a spiritual sense we think of good and evil as being light and dark. Goodness - or Godliness - is "being in the light." That we are using light as the medium for the connections we enjoy through the Web becomes a fascinating metaphor for our greater humanity.

If you picked this book up wondering whom it was written for, the short list is: moms, dads, teachers, preachers, doctors, government officials, students, business people and technologists. In other words, it was written for anyone who is curious about how technology might be impacting us and how it is likely to impact us in the years to come. If you've ever been curious about the long-term impacts on our kids, what it will do to our brains, or asked if our lives will really feel better when we are more highly integrated with digital tools, you are in the right place.

I have been in the technology field for over thirty years and I love the transformational capabilities software, hardware, and devices give us. That means I got started when personal computers were a new concept and a floppy drive was the hottest new feature in storage. I am a husband, father, grandfather, soccer coach, business builder and a pretty normal human being. I believe in God, the existence of good and evil and, more specifically that there is an energy of love (light) in the world that supports goodness. I also believe that there is energy of evil (darkness) that acts to destroy. These factors have combined in my life to make me very curious about whether the darkness or the light is slowly winning? Are we becoming more enlightened as a species or more corrupt? Or, does there exist a "perfect" balance where we can never escape an equal amount of good and evil? And particularly relevant to this book: will technology impact this balance in any way?

For many years, I have been intrigued with the role technology plays in shaping who we are and how we operate as human beings. The blessing of being my age is that I have been able to watch the impact from the PC era to our current immersive, digitally connected world. I grew up without much technology in my young life and now I spend most of my time consulting with companies on how to have technology mastery. I have seen both sides, and that helps me have a balanced view of some of the positives and negatives of the impacts we are now experiencing.

When I give speeches (which I do for a living) I am often asked how I feel about the impact of technology on our children's privacy, our stress levels, the economy, and our long term happiness. The more people ask these types of questions, the more I became motivated to write a book that starts a discussion about how technology will impact the human race. Many of us are very curious about this, yet it seems to be a subject that lacks a holistic viewpoint.

So why me? Why now?

I am intensely interested in the future because I have such love for my family and the generations that are already being born. I also care deeply about you and your family. Because we choose thousands of times every day to align with helpfulness or harmfulness, I want to help us all see that technology amplifies these choices tremendously. We are inventing many different ways at a rapid rate to augment our lives digitally, along with our bodies and brains. If collectively we make poor choices over the next few decades, the consequences will be devastating.

I have read this statement in articles or books many times: "The seeds of the future are already planted today." If we are to manage how these seeds grow and blossom, we must first gain a better understanding of our potential future in a highly technology-augmented world. We must be able to turn on our intellectual high-beams and have an understanding of how the inventions of today will impact us tomorrow. I have painted a number of pictures of the future for you in this book and dearly hope they give you a means to make better decisions today.

Trying to predict the future is a discouraging and hazardous occupation, because the prophet invariably falls between two stools.

• *If his prediction sounds at all reasonable, you can be quite sure that in 20 or at the most 50 years the progress of science and technology made him seem ridiculously conservative.*

• *If the prophet could describe the future exactly as it was going to take place, his predictions would sound so absurd, so far-fetched, that everybody would laugh him to scorn.*

If what I say now seems to be very reasonable, then I'll fail completely. Only if what I tell you appears absolutely unbelievable have you any chance of visualizing the future as it really will happen.

Sir Arthur C. Clarke, 1964

Although technology is the most important character in the story, this is not a technology book. Although it has "God" in the title, it is not a "Christian" book. This is a book about our human journey and what the impacts are likely to be of a powerful set of new tools. Now we have powerful streams of information and vast problem-solving capabilities in our hands every moment of every day. We each have a soul also, and these two statements actually impact each other. Whether you agree or disagree with my point of view, this discussion needs to be had. We must be more proactive in thinking through where the technology path we are on will take us a few generations down the road. Short-term thinking with some of our inventions has caused us much pain as we one day learn they had very negative long term impacts.

I have a friend who goes away to a cabin for a weekend at least once a year to think big thoughts. I respect him for taking time from his busy career to do this. I was visiting with him one day after he had been on one of his trips and I asked him what he had worked on this time. He explained that he has a 200-year plan for his family and that he was looking at what he needs to do now in order to help his family tree most through at least a few generations. I was impressed with his foresight

and willingness to look down the road that far and to care about future members of his family that he would never know. I asked him a lot of questions about what he was doing differently now in order to impact his family later and he gave me an admixture of financial moves and spiritual maturity. By providing for their physical needs through the building of generational wealth, he planned to help them have lifestyle options and opportunities for a broad range of experiences. By providing a great example spiritually and teaching them the importance of loving people, he planned to help them have a more peaceful existence. You have to admit his 200-year plan is an impressive commitment!

Did God create the Internet? The answer to this question has profound implications for the future of humanity. Will we become androids or cyborgs devoid of humanness? Or will we retreat from technology and hide from its benefits, as well as its challenges? How will our attachment to technology affect our lives long term – will it pull us towards earthly things and away from a desire for enlightenment? Or will it play a central role in our awakening to new levels of enlightenment? Will we develop the wisdom to find a harmonizing balance between the unique gifts we have as human being and the tools we invent to augment how we operate on this planet?

This book is my attempt to address these key questions. My hope is that, after reading this book, you will be able to answer these for yourself. It is not my place to tell you whether God created the Internet, or even whether there is a God. Each one of us has to answer this question for himself or herself.

When I say that our future depends on how well we learn to integrate with technology I am not really saying anything too earth-shaking. Already we have learned how much technology changes our world. Whether you believe in God or not, it is quite obvious that we have free will. We will all choose collectively the impact technology has on us. If you do not believe in a spiritual creator, the question still remains: What will happen as the balance between human traits and technological capabilities tilts? Another reason to answer the question stated in the title is because I believe strongly that technology is giving us the chance to connect, learn, share, and enlighten our human race. We have never had a tool as powerful as the Internet, with the capability to connect people instantly, anywhere on the planet. Maybe we were given this tool for a specific reason.

Much of what I will share with you is a combination of technology, psychology, philosophy, spirituality and future vision-casting, all of which might seem like a strange brew. But all of these impact our lives on a moment-to-moment basis and the future impact will reorder how humankind exists. There is a very real chance that we will integrate with technology in ways that literally evolve us into a different species (transhumanism). As we progress on the journey of integrating technology into our lives, it will increasingly impact our minds and emotions. This is both breathtaking in its implications and sobering in the potential outcomes for our race.

I am sure there are many people who believe there is no relationship between technology and spirituality. Let me give you a simple concept that might illustrate the connection in a clear way. The Internet is a human-constructed entity that connects devices and, ultimately, people. It is web-like in structure, which of course is how it became known as the World Wide Web. This structure is providing a historically significant ability to store and forward information, and to connect everyone who are on it in real-time. In just a couple of decades it has provided us a way to connect billions of people in a way we have never seen in our history. God clearly wants us to be One. In order to live out holy principles, every person in the would would need to connect with each other in loving and accepting ways. We would need to exist with each other in a frictionless way, where we all are different in skills, talents and gifts, and at the same time are "one body." If we take these two different ideas (the Web and Oneness) and think of the Internet as a physical step that helps us move toward Oneness, then we can see where technology and spirituality might overlay.

As to the spiritual impacts, does technology help or hinder spiritual growth, and does spirituality help or hinder technological growth? On one side there are highly spiritual people who believe that technology is the antithesis of a sacred spirit, and that it does nothing but blind us to goodness. On the other side of the spectrum there are people who do not believe there is a spiritual element to our lives. I will do my best in this book to provide data points along the way that will help you see the potential positive and negative impacts of technology on our spiritual lives. If you are uncomfortable with spiritual principles, then you can just think in terms of what technology does that is helpful and unhelpful.

Before going any further, let me define a couple of terms used in the book. When I talk about technology, I am generally referring to Computer and Internet-based technologies. I realize that the word "technology" can be used in many ways, and to describe many different types of new inventions (biotech and nanotech for example). So that I do not have to write the phrase "computer-based technology", or "information technology" over and over, let's just agree that I am referring to computer and online technologies when I use this word. There is a glossary at the end of the book where I have defined terms that are technology-related and may not be generally used. Please refer to the glossary anytime you run across a specific word that is not familiar to you.

Humalogy is a word that we crafted at Future Point of View (our company) to describe the integration of humans and technology to complete a process or task. Most things we do today are a combination of both, so we can think in terms of a blending that could be measured using the following scale:

	H5	H4	H3	H2	H1	0	T1	T2	T3	T4	T5	

Zero on the Humalogy Scale represents a perfect balance between the use of technology and humanity to complete a task. As an example, think about driving a car. This requires a person to choose the destination, to push down on the gas pedal, to steer and to guide the technology that is the car. There is an integration of a human and technology in order to achieve the process of driving. This same concept can be applied to how we find a job, manufacture a product, build a house, shop online and communicate with people all day. The process of giving a baby a bath today is normally an H4.5 because we use very little technology to accomplish this. The process of buying products online would be a T4.5 because we rarely connect with a human being when we order online. In each case there is a percentage of the task done with our hands and minds, and a percentage done by using technology.

As a task depends more heavily on technology it moves through T1 all the way to T5 if it is completely done with digital tools. H1 through H5 measure how much human involvement in the task. The Humalogy Scale gives us a way to denote how the combination of human and technology efforts might be shifting over time. It also gives us a way to understand the balance between the two so we can debate what might be the most appropriate or efficient balance.

We deal with Humalogy in most parts of our lives today without really being conscious of this balance. For example, think about how deeply technology is ingrained in how we do many tasks in our our jobs. Watch a gamer, deep into a virtual environment, controlling an avatar through a collection of devices and software You are seeing that entertainment has a Humalogical balance. What about computer-assisted surgeries? Or the modern-day excavator or backhoe? These examples are just the crude beginnings of how we will integrate our bodies with technology to fulfill just about every commercial task we do on a daily basis. People integrate with these machines to the point that they feel like an extension of their own bodies.

Obviously the balance is shifting on our scale. We are moving inexorably from the "H" side to the "T" side for many of the activities in our lives. This is providing for us wonderful capabilities and efficiency. It also comes with potentially negative consequences. The possibility exists that, when we move the balance too far toward technology, we can lose the great value of human-provided gifts. It is not always better for the world to erase humanity from a process. For this reason, the ability to find the perfect Humalogical balance for any process, or for our lives in general, is critical. To be out of balance will have grave consequences for our happiness and general well being. As you go about your day today, use the concept of Humalogy to discern how much technology you are using to accomplish tasks and see if you have created the best balance to accomplish each one. Could you talk to a person face-to–face instead of texting them? Would it make sense to take a long walk with a friend instead of watching a show online? Too often we are unconscious of the Humalogical balance and that being too far to one side or the other has the potential to create problems.

There is another concept I will use a number of times in the book: the idea of the "Universal Mind." In the technology world we will sometimes use the term "Hive Mind." In both cases the meaning is a collective ability to share thoughts, information and questions across a large number of people. Today we are using the Internet to do this, and we are creating a world-wide aggregated level of intelligence we have never had before. Our minds might have been shared in small groups all through history, but we have never been able to store information and share in real time the way that we can today. We truly are creating a hive/universal mind that is an entity all on its own. As we do this we are also stepping toward Oneness because we are sharing with each other at unprecedented levels and once again this highlights the overlapping of technology and spirituality. It is this concept of Oneness that leads me toward using the term universal mind, which I think better describes where I hope we end up. Universal mind has a connotation of being constructed by mindless drones and I like to think humanity is more than this.

This is no small tool for humanity. We all know what can be accomplished when we apply thousands of people at one time to do work with their hands. We built the Great Pyramids, we built the Great Wall of China, and we built the Hoover Dam. In physical, human terms, the power of aggregating people is clear. Now we are moving into an era where we are combining our minds. This is much more impactful than combining our hands and backs. The Internet has provided the connection to allow us to combine our minds into a collective knowledge base, through saving information on servers and hard drives for instant access. It is also providing a real-time capability to share ideas, solve problems and communicate with each other. Most people do not see the Web as facilitating the universal mind; they simply recognize it in the day-to-day capabilities they use. It is important to step back and realize that the Internet is more than just a collection of applications we use each day. It is the great connector of our minds, and this will dramatically change our world.

Which takes me back to the central question of this book: Did God create the Internet? This question has been banging around in my mind for many years, almost since the Web was first made available to us. I believe we were created by a higher power that gave us free will to choose how we use our energy. I also believe that God loves when we use new tools to broadcast and amplify the good, and mourns when we use the same awesome power to dive into the dark side. As humanity invents increasingly powerful tools of connection, automation and calculation, will we choose to use all of this technology to elevate the human race? Or will we choose paths that devastate us over time?

In order to start this discussion, we must consider the positive and negative aspects of what technology is doing to us today and then move on to what is possible in the future. I will do my best to present observations and facts without adding too much personal bias so that you can make your own decisions about where Humalogy will take us. We now find ourselves stumbling forward at an intersecting road of hard, cold silicon, software, humanity and spirituality. Unless you plan to disconnect completely from society, you will be greatly impacted by this new path, so it would be good for you to have your eyes wide open.

I am very certain that technology is just a tool – like fire or guns – and it can be used for healthy or unhealthy purposes, depending on the person wielding it. We live in a world that is immersed in both good and evil, so there is no reason to think that our use of technology would be anything other than a representation of how we all choose to live. Humans seem to have a propensity to explore the full range of use for any new tools. The question is: will there be a slight lean toward enlightened use, or an alternate path toward disconnecting and destroying our human gifts? We cannot know for sure what the future holds; certainly we can influence it if we choose to take the responsibility to look 200 years down the road.

CHAPTER ONE:
Standing at the Crossroads

Technology is a way of organizing the universe so that man doesn't have to experience it.

—Max Frisch

I strongly believe that at this moment in history we are at an inflection point in our progress as a species. We are standing at a crossroads of sorts where the steps forward over the next two decades will dictate heavily the quality of life for billions of us. Historians will look back for hundreds of years and judge us based on our collective handling of the digital transformation we find ourselves in. Digital tools and their uses are having a massive impact on how we exist, and this will continue changing rapidly how we work, play, and connect with each other. The technological choices we make over the next decade or so will forever define us as a species. Most people don't recognize that we are dealing with not only a historically significant time, but also an era where we will be offered choices that will dramatically impact our species. Many of us do not consciously choose our road forward; we blindly follow the herd, or resist making any choice out of fear of change. We spend most of our energy surviving and too little thinking about the future we will live in. This has led to a long history of being ignorant about the impacts of our inventions, at least until the impacts are felt personally and, sometimes, painfully.

Here are a few examples of the trouble we've gotten into when following the herd with our technologies:

• We loved the results of using pesticides (DDT) until it got into our food and made us sick

• We love our vehicles and have built a culture of driving even while auto accidents have become one of the leading causes of death

• We love television and it often makes us less active, less healthy, and exposes young people to content that is not always positive

• We love processed food because it is convenient, but it causes us to gain weight and to be sick at times

• We are in love with the way alcohol and tobacco products make us feel regardless of the fact that our abuse of them has maimed or killed millions

• We invented and fell in love with a host of legal and illegal drugs that have destroyed the lives of users and their families

• We enjoy immensely the openness and massive amount of content that anyone can post on the Web and because of this capability we grew our addiction to porn

• We love texting and now thousands of people die each year from distracted driving

We are now falling more entranced with every new capability digital tools can do for us. Let's learn from the past and be more careful to understand the consequences this time around.

For decades people have been trying to envision what the future world will look like as humans and machines continue to blend. For some, their truth is illustrated in dystopian movies like *The Matrix* and *Terminator:* a world where machines have taken over and rule us. We spin tales about machines developing their own personalities and deciding to do us harm, as in the movies iRobot, and Hal from the Space Odyssey series. The underlying theme seems to be that, at some point, technology becomes self-aware, and when that happens, it will exhibit the human qualities of self-preservation, or a soul-extinguishing greed for growth. It seems clear from the stories we are telling ourselves that we have a lot of fear about where technology will go in our lives. Here is what I have observed from the people I engage with while giving speeches:

Top Ten Technology Fears People Have Shared with Me

1. We will be addicted, and then crippled, by using so much technology

2. We will become dumber because we'll let the technology do so much thinking for us (or solving so many or our problems for us)

3. It will cripple our ability to have face-to-face connections, write well, have social skills and will disconnect us from family and friends in some ways

4. It will take away all of our privacy and make our lives transparent to all, whether we like this or not

5. It will cause all of us to have attention deficit problems because we are always connecting to a device and its river of information

6. We will never be able to get off the grid, get away from work, be free of digital connections, and to know the value of silence

7. Technology will replace humans and their jobs, leaving people unemployed or underemployed

8. This could lead to class warfare, as there will be an upper class of technology "haves" and a lower class of technology "have-nots"

9. Eventually technology will control us and tell us what to do from moment to moment, instead of the other way around

10. It is getting so complicated and there are so many devices and apps that we will never be able to keep up

We must be aware of these concerns, and at the same time, not assume they are a foregone conclusion. We must acknowledge that we are integrating technology into our lives and bodies to augment ourselves so that we are more loved, productive, and better entertained. Our motivations for shifting the Humalogical Balance are not always wrong. Every day there is a new technology-based concept being birthed, and this will not stop any time soon. There is nothing wrong with that as long as what we develop has a healthy purpose and is used while balancing the gifts of humanity with the efficiency of technology.

Some of the stories we write about the future postulate that many of us will tire of technology and go back to a place where we live simply, "like we did in the good old days." I find this to be thinking driven by a need to go back to what is familiar. At the same time, I will never judge someone harshly for wanting to disengage with technology completely. I do predict that, other than an occasional "off the grid" vacation, most people will not want to disconnect fully. The truth is that the vast majority of us love what technology does for us. It makes us more productive, it connects us to people and information in amazing ways, and it entertains us. In short, we believe that it improves our quality of life, not that life cannot be wonderful without all our gadgets. There are moments in life where technology plays a wonderfully positive role that has nothing to do with actually using it.

A few Christmases ago I watched a wonderful scene unfold involving three young girls, their iPads and a tragedy. They were thirteen at the time, and one happened to be my daughter, Kristin. She had two friends from school, Hailey and Caroline, and all three played soccer on the team that I coached. Every good story has a villain, an overcomer, and friends along the way. In this case, the villain was cancer. Caroline was the overcomer, and Kristin and Hailey were the friends. Caroline's mom died right before Christmas and that was devastating to her, as you can imagine. Caroline is, to this day, a wonderful person. As a younger kid she was always joyful. Seeing her in such pain was difficult for anyone who knew her well, and especially for her two friends who were too young to know how to say the right words or make her pain go away. Even though she watched her mom go downhill for many months, there is a brutal finality to a parent dying. It leaves a huge hole that cannot easily be filled. This was written all over her face, as she was much too young to lose her mom.

Kristin and Hailey were both getting iPads for Christmas, and at that time, this was the latest and coolest technology around. As soon as they realized they both would have this gift, and that Caroline would not be getting anything like this for Christmas, they hatched a plan. They came to their parents and asked if there was anything they could do to earn the money to buy an iPad for Caroline; they simply could not live in a world where their friend would not have the same device that they had. Hailey's parents and my wife and I were touched by their fierce desire to do anything they could to help Caroline, and in truth, we were heartbroken for her family in general. Her mom was a wonderful person who was cheery to the end and could not have set a better example of dignity while dealing with cancer.

To this day, it is hard for me to tell this story without tearing up because I remember in detail the fact that the girls literally cleaned our bathrooms to earn money: something they had probably never done before or since. They would have done anything we asked to help Caroline have a better Christmas and to help her share in the fun of having an iPad. We made them do the work for three days solid and they never complained or shirked anything we asked. We would have bought the iPad for Caroline in a second anyway, but we wanted them to feel that they earned it. So we paid them ten or twenty dollars at a time until they had what they needed.

When they gave it to her, there were many tears of course, because it was not about the iPad. It was about connection, and to their generation, technology facilitates being able to connect to each other at a moment's notice. It represented access to information that would help her get homework done easier. The world was a better place because Caroline got what they got for Christmas.

An iPad does not replace a mother lost. What it did show Caroline was how much Kristin and Hailey loved her and now they had one more tool to bond them together. That was the best way that two thirteen-year-olds could have imagined to say "we love you" to a friend who was hurting.

Ray Kurzweil talks about what he calls The Singularity, his theory that, at some point in the future, computer systems will cross over to being able to process information just like, and then better than, the human brain. At that point, it is reasonable to assume computers will have the ability to learn on their own. Because their scale and ability to digest information will be much greater than ours, technology will move rapidly past us to a point where it will be able to solve problems we cannot even imagine. He predicts this will happen in the middle of this century. Even if he is off by a few decades, many people reading this will see that day. If we can accept that a machine one day will be able to process information in the same ways we do, and that it may even develop its own point of view based on what it stores in its databases, then we need to be thoughtful about how to harness our computing power for the positive, and guard fiercely against the negative.

I believe that all the post-apocalyptic movies and books we generate are a siren song to ourselves about the consequences of not being thoughtful about controlling our computing power. Somehow we know intuitively that at our present rate of progress we will certainly hit The Singularity at some point, and that this tipping point must be guided by our maturity in preparing laws, social mores, and digital boundaries that keep our technology from going rogue, intentionally or not. I have been heartened to read about some of our thought leaders in society talking about practical solutions, like a worldwide agreement to not build self-targeting drone-like weapons. It just makes sense not to build an unmanned weapon that chooses its own targets of opportunity, because the slightest change in programming would create a machine that blindly seeks to kill whatever crosses its path. Ultimately what we need is spiritual maturity so that we do not even need these kinds of laws to tell us to protect ourselves through rules. We need to protect all of humankind because we choose to honor all life.

It may not matter whether computers can surpass the brains ability to process independently because there is no question that computers will augment our brains. In fact, they do already. We know technology will augment many aspects of how our bodies works from now on; we have already crossed the line of combining human and machine. We do it every day with mobile devices giving us instant access to the Internet, and with every disabled person who is helped with prosthetics. Throw in decision support software that helps make business decisions for us and we can begin to understand that we already have the first steps in place toward turning ourselves into the human/machine hybrids shown in science fiction. We scoff today at the thought of people wanting to be cyborgs, yet it is not a preposterous thought if we really admit our attraction toward augmenting our capabilities to give us an advantage in life.

From a psychological viewpoint, it is interesting that virtually all our stories about the future are dystopic. There is a completely different picture we could be painting and have not yet, however: one in which technology helps create a sort of utopia for us. Every once in a while in science fiction there will be a planet or floating platform that houses an advanced society apparently positively impacted by technology described in the story. In almost every case that society is attacked or destroyed by the "evil" force wielding a technology-based inventory of weapons. Maybe this is the human fate we are saddled with for the rest of our species days: the fight between good and evil, even when we are highly technologically augmented. I certainly hope it is not so!

Maybe the reason we often see such a dark future can be explained partially by the fact that we seem to find negative things to do with technology before positive uses take over. A great example of this was the growth of the online porn industry for the first decade that we had the Web. It took many years before social technologies became the more heavily-used applications supplanting the porn business. To this day, the ability to see pictures and video of any sex act is so widespread it is impossible to keep our ten-year-olds from seeing visuals that would break our hearts. It seems like humanity often finds harmful uses for any tool we invent before finding the helpful. With better future vision and intentions, we could skip the early unhealthy uses of technology and instead go straight to things that bring goodness to the planet!

Technology is a giant toolbox that humans will wield for good or evil. How enlightened we become as a species will dictate our future ability to survive and prosper as these tools become immersive in our lives. How enlightened we are depends heavily on what our beliefs are about why we are here, what we are meant to be, and what we are meant to create on this Earth. Which brings us back to the integration of Digital and Spiritual. When a person has a gun in their hand, their level of maturity will dictate how that gun is used. When a leader has a nuclear warhead under their control, their beliefs about the fundamental nature of humanity becomes critical to how they might use the weapon, or choose to dismantle it. In other words, where lives are involved, we all hope the person with the tool in their hands has positive intentions.

I do not believe that technology, on its own, will seek to dominate the human race. I do believe that software we build with self-learning systems will aggregate huge amounts of information and come to wrong conclusions in some cases, just like humans do at times. I also believe that we will build robots that will have programming flaws and they will kill humans without any intention of doing so. Then we will fix this problem so it does not happen again. We will certainly build knowledge-based systems that are smarter than any one human; we already have this in specific areas of knowledge as IBM's Watson (their supercomputer) has proven. I do not believe that a machine ever will have a soul, but I do believe that a machine one day will desire to protect itself to the degree that it will kill humans in order to "survive." And I believe it will only do that because someone will have programmed it with cognitive capabilities so that it has some ability to think on its own. Then, as a society, we will have to decide how we will hold the person or company accountable for allowing the technology to go rogue.

We are certainly at a crossroads in our journey forward. We stepped into a new era when we built a method for connecting all of us and storing our collective knowledge. We are enhancing the Internet all the time by building devices that move it closer to our bodies. We will, from this point forward, always deal with finding the appropriate Humalogical balance in everything we do. This will ripple through the fabric of humanity at every level. What I seek is to make sure that as many people as possible go forward into the digital transformation with eyes wide open. There are moral/ethical obligations that could be woven into how we approach technology and the more we would be willing to accept obligations like these the more likely we will have a positive future. As examples, I have created a short list of obligations broken down into personal, organizational, and societal categories. As you read through these try to imagine the impact these would have if we would be willing to commit to them!

Personal Obligations

- To balance time spent interfacing with technology and time spent engaged with people

- To consider carefully the content and tone of content posted or re-posted publicly

- To not use technology to further an unhealthy addiction

- To recognize when the use of technology is taking our attention away from critical tasks that require our focus e.g. driving, listening to others, walking, or participating in activities

- To not allow our use of technology to impact negatively those in our immediate surroundings

- If we provide a child with access to technology, we must then take the responsibility of helping them learn use technology in a healthy way

- To be honest when we create online profiles and use avatars to connect with people virtually

- To treat other people's devices, data, and files as private property

Organizational Obligations

- To fully train people to use the technology tools required for their positions

- To handle displacement as humanly as possible – this includes adequate notice and possible retraining or repositioning

- To consider the physical implications caused by increased time at a device

- To draw boundaries between work hours and off-of-work hours

- To consider privacy concerns in the areas of data and activities

- To consider the environmental impact of a technology decision

- To consider the future dangers of the use of their technology products or tools

- To not use unethical techniques to besmirch a competitor's reputation online

- To be fair and ethical with all digital marketing techniques

- To consider customer's data as sacrosanct and not share it without permission or abuse the use of it

Societal Obligations

- To project how the technology being invented might impact the human race in negative ways and make wise decisions as to whether to build it all, or to at least build controls to that assure a positive outcome

- To provide access to key technology tools and capabilities to all people as a matter of human rights

- To be unbiased against people who choose to opt out of using technology if the motivation is to gain simplicity and peace in life

• To provide training and design capabilities that allow people who may not be comfortable with new technologies a path towards learning so they are not disenfranchised

• To provide dependable and clear levels of privacy in the areas of body and activity privacy and some level of control over information privacy

• To provide a consistent and high level of law enforcement over the Internet and cyber crime

• To constantly improve the digital tools we are building so they honor spiritual principles and help to improve the enlightenment level and prosperity of the human race

At the end of every chapter I will summarize the core thought for that chapter and include a list of seminal questions. These are questions for consideration and contemplation. Please think about these questions and ask yourself how they apply to your life.

SUMMARY:
Standing at the Crossroads

There are moments in history where humankind will move in a completely new direction based on the new tools we invent. We can look back and see how the wheel, the ability to create and shape metals, build boats and build new weapons all changed the course of history. As the world became more populated the overall change of inventions like cars, airplanes, telephones, television and advanced medicines became massive. As we were able to extend the life expectancy of people, the world became more populated; that has huge implications. As we provided better ways to move people around and allow them to communicate, the world became a smaller place and people spread out around the globe. New inventions make a big difference in how humanity progresses.

Computers and the Internet one day will be looked at as one of the more impactful catalysts in our history. When the smoke clears from the impacts technology will bring, we will look back and hardly recognize life before having these tools. History before the Internet truly will feel like the Dark Ages. This is the reason we need to understand that we are at a crossroads in history. The better visibility we have into the future impact of our technology, and the more we seek to have healthy outcomes, the better life will be for all of us. The more we go blindly into this very changed world, the more we will suffer the negative consequences that come from being out of Humalogical balance.

Seminal Questions

Is the pace of technology innovation going to speed up or slow down in the future?

In a world of good and evil, which side will have more success using technology as a tool?

Will we experience more love, joy, and peace as we become more technologically augmented?

Will the integration of technology into our lives and bodies make us more or less human?

Will technology ultimately move us forward as a species or destroy us? (Utopia or Dystopia)

CHAPTER TWO:
Technology, Outcomes, and Choices

1 in 3 people consider the Internet to be as important as air, water, food and shelter.

—Source: psychologytoday.com

Inventions of all sorts have been changing humanity for thousands of years. The only difference today is the scale and velocity at which those changes come. The telephone in the 1800s was a huge step forward, yet it took decades to connect a small percentage of the people on the planet. Compare that to the speed of mobile device usage which went from the bag phone (the most recognized early version of a mobile phone) to the iPhone in twenty years, and from zero users to billions in the same timeframe. This speed will only increase and I am sure that in another twenty years there will be an online/mobile application that will hit the market and be downloaded by over half of the people in the world in 24 hours.

Although that does not sound odd to us today, the thought that four billion people might adopt a new tool in one day is staggering, especially if that tool changes how we learn, communicate or function in some important way. Can you imagine two billion people all over the world getting the ability to talk on the phone at the same time in 1860? How might things be different today?

Television Conquers Our Homes
For our purposes, a good place to start with this examination of technological influence on society is the television. The advent of television brought the first ability to transmit video over long distances. Visuals, as we have learned, have a

huge impact on us. Television began its march to proliferation in the 1950s. The numbers of homes owning a television set increased rapidly in this decade, from 0.4% in 1948 to 83.4% in 1958. That is an amazing statistic for technology adoption at that time in history. In just ten years, television became mainstream even though it was not an inexpensive device. This rate of adoption for an electronic content delivery device foreshadowed the quick pace we have now seen with computing devices.

I was born in the early 60s and as a young child had four channels on a black and white TV. I do not remember life before TV, although there are certainly still people alive who do. As a child, most everyone I knew already had a television because I grew up in a suburb where most people could afford one. By and large it was only the few households where a parent was diametrically opposed to people sitting on the couch for hours that eschewed this device.

The impact on humanity from what television has delivered into our homes is staggering. For all of history up until the 50s the only visuals a person had of the world came from the reality around them, pictures in books, and eventually movies in theaters. That all changed in the 50s and 60s when households could gather around the one TV and watch a baseball game being played across the country, humanity landing on the moon, or the funeral for President Kennedy. From an entertainment standpoint, this was a watershed moment for the world. From a historical standpoint a line had been crossed, because information could then be beamed all over the world. Of course that also meant whoever had control of that information was powerful because of the political, moral and intellectual impact of influencing people by the millions. And with that control of television content, what did humanity do, this story is one example:

In 1988 I began working in the U.S.S.R. to start a technology company. Having not seen much of the world at my young age, I was not prepared for the vast differences between the world I had grown up in and the world their citizens endured. Not only was their quality of life substantially below ours in the U.S.; their knowledge of the rest of the world was stunted. People today do not think much about the concept of the "Iron Curtain" but it was something I got to see firsthand.

The Soviets had three television channels from what I recall, and all of them were state-run. That meant they had a very restricted window on the world compared to what we had. The result of this was that the Communist party was able to convince just about all of the Soviet people that the rest of the world was in worse shape than the USSR.

While I worked there between 1988 and 1992, I met many people who for the first time were allowed to go to the West through business, sports, or the arts. I will never forget one woman who, upon returning from a short trip to the U.S., found herself very disillusioned. For over forty years she believed what the Communist leaders had told her about the lifestyle of the West. Then she walked through one of our grocery stores and was stunned by how many products we had and how clean everything was. She walked the streets in the city and found that they were not lined with homeless people as she had been led to believe. Such are the results of blocking, restricting or controlling a technology like television.

In the U.S., sometimes we feel like the television delivers too much information. I saw firsthand what happens when it did not deliver enough. The Internet is now posing the same kinds of problems. Today we have countries and leaders who try to control the Web either because it delivers information they deem morally bankrupt or because they really do not want their citizens to connect to the rest of the world. The Internet is easier to deliver into an area even when the leaders might want to restrict it and in this way, it provides a critical service to the world. The free flow of information is important to providing transparency about the truth and at the scale of an entire region with millions of people, transparency if important to verifying basic human rights.

There was (and is) more going on than just entertainment on the magic box. Television helps us learn new things about the world, and at a faster rate than we had previously gained through books and personal observations of the world within eyesight. If a picture is a thousand words, then hours of video delivered to your home is worth millions of words. What we see with our eyes and how that imprints on our brains alters us in many tiny ways. In some cases, it effects major life-altering

33

swings. The impact of one show that millions watch regularly is powerful, whether positively or negatively. Today, a viral video on YouTube can do the same. What a young child sees on a television screen forms for them what is acceptable or not in society, and in ways that might be very different from the wishes of their parents.

In the 60s this was not a huge problem because most of the content delivered into the home was pretty benign and many families watched shows together so the parents could easily control what was viewed. In the 70s and 80s that began to change as sitcoms became increasingly edgy. We moved from The Cleaver family and Dr. Marcus Welby to the Bunkers and the Jeffersons, and this changed the tone from programs simply offering family entertainment; now they were political statements. The shows mirrored the political climate in some ways, and in others, they directed us by creating new viewpoints of the world. What a few people created on a set in New York or California impacted millions of people around the country by mixing increasingly edgy viewpoints into our entertainment. As the 80s progressed there were many shows added that were not the kind of wholesome entertainment we saw in the 50s and 60s. This new breed of content was meant to educate or inflame us. We were meant to be shocked by immoral or even criminal behavior we were witnessing. Although we had always had crime and evil in our midst, the world just seemed to be less safe when we saw the stories all day long.

By the 90s producers of shows had learned that the more one pushed the boundaries of norms, the more buzz they got and the more viewers they were able to attract. Deliver a little nudity or bad language and people were intrigued. Mix in showing alternative lifestyles that the public was not generally familiar with and even more curiosity might be generated. Add a little bloody violence to the mix and some people were even more attracted. Eventually we created reality shows that allow us to be voyeurs into other peoples' lives, and we become attracted to whoever could behave the craziest. The final ingredient for capturing peoples' attention was to show someone famous behaving badly or being destroyed in some way. This always gets people to tune in. Show producers only had to be careful to not step too far over the line of decency or advertisers would go from being supportive to pulling their sponsorship if a public backlash reflected on the advertiser's brand. The sponsors might bail and that means profits drop, and then the content is not financially feasible to deliver.

At the same time we also developed shows that explored all corners of the world. Producers used their larger budgets to fly people to every interesting place, interviewing all kinds of people. This helped many of us learn fantastic things about parts of the world we may never see in person. It broke down barriers because we learned that, although people might dress differently and speak different languages, they were very much like us in many ways too. We saw beautiful geography, sports we had never seen and learned about customs of people halfway around the world. This is an awesome gift that television gives to us: a glimpse into worlds most of us never have, and never will, see in person.

Over forty years we went from wholesome and relatively healthy TV programming to having a number of shows that were titillating, salacious, violent, or out of the norm. We also gained windows on the world we had never had before. Understand that I am not prudish about all the changes in content, and I lived through watching this entire transition. I am just stating what happened, and it was a dramatic shift over a couple of generations. Producers and networks who make money by supplying content that people will watch (which makes advertisers happy) always are searching for something that will attract eyeballs. They experiment with every kind of content and stick with what works. The reason we have such a huge mix today on our 500 channels is that TV now reaches a multitude of people with different interests, and frankly because we can provide hundreds of channels!

Today there is very little we would not broadcast because it is possible to see just about everything that humankind engages in on cable. When a young child grows up seeing risky or ugly types of behavior on the screen, he or she sometimes sees this content as acceptable, even when what it is clearly unhealthy. Add this to the fact that violent programming slowly numbs people to the realities of true life violence and risky sexual behaviors, and it is easy to see the negative aspects of putting pictures in viewers' minds. Did show producers simply give us what we wanted, or did they entice us into watching content that was edgy and shocking? There is no purpose in placing blame because the important point simply is to see that a technology (television) has had a life-changing impact on humanity.

To be sure, there are many wonderful things about television programming. Aside from the pure entertainment value it brings, it allows candidates running for office to be exposed to millions of people so that we have a better idea of what they stand for. There are channels today that teach us how to improve our cooking, how to exercise, how to build things, and how to do just about any task to improve our homes. The news informs us of the larger world we would otherwise not know anything about. Investigative reporters hold people accountable to do what they say they will do. If someone is dishonest or shows lack of integrity, networks are more than happy to expose this, even if that someone is a President of the United States and/or the most powerful person in the world.

Ask one hundred people who their hero is and you will find that many of them will name someone they grew to know through TV. Studies still show that we spend many hours a week taking in content from a television. The shows we watch can come from anywhere in the world and are often shown live. Without television, we would have a much more limited view of the world and the people in it. It is simply the most powerful tool for connecting us to the world in ways we could not have imagined back in the 50s.

Television and the Web

It is not only the content we are exposed to that influences us: advertising does as well. There is an analogy between TV and Web advertising impacts because they both require an investment from us. With television we trade our time by watching commercials and are influenced to spend more by advertisers than we should. This is the price we pay for entertainment and, although annoying, it is much less invasive than what we now pay for online services. In the digital world we are able to download and use many free services online because they are advertiser-supported. That means that either we are inundated with banners, interstitial ads or we give up our personal data as the payment for free applications and services. The conscious - or with some people, unconscious - trade we make is to give up a large part of our privacy while websites harvest our personal data. This is the price we pay for online services.

By the time a child spends 18 years watching what advertisers and network producers feed them, they have formed a view of the world outside their homes that is a blend of what parents and educators have tried to ingrain and what strangers have delivered. Depending on how that mix was digested by the child, habits are formed: some positive and others, devastating.

There is another interesting analogy between television and modern computers. In the computer world, we have a device (a PC), a transport system (the Internet), and software (the applications). We have roughly the same with television in that we have a device, the transport system (cable or wireless, so to speak), and the programming (the software applications). Over time, all three changed dramatically, in that we went from huge wooden boxes with tiny round screens of fuzzy black and white to huge flat panels with vibrant color. We went from big antennas to high-speed cables. We had three main channels; now we have hundreds. They used to deliver primarily family entertainment, and now we have pay-per-view porn channels and reality TV. With both television and the Web, we have seen the devices and delivery systems constantly improve and the content grow in volume. All of this has expanded our horizons with its breadth and depth.

There are other parallels between TV and the Web. When the Internet first started, it was just a transport protocol for email and text-based information. With the advent of the World Wide Web, we were able to add graphics and sound, and then we added video quickly afterward. For the most part the Websites that were brought online were for the purposes of organizations connecting their information to the public. More quickly than with TV, we added porn sites, and once again a young child had a window on the world that was expanded and set for them what they perceive as acceptable.

More than 58% of children surveyed (ages 14 -17) report having seen a pornographic site on the Internet or on their phone.

Source: parentstv.org

Hollywood and the television networks will argue forever that they simply gave us what we wanted and what we needed to be entertained. Website providers now argue the same thing. They will say they simply experimented with content and we chose what was popular. They will avoid accepting that they are slowly corrupting our collective moral base. They will say that we have always had moral issues in the world and that their content delivers what we want. It is a chicken-or-egg debate to be sure. Did television change us or did we drive television content by what we gravitated toward watching? Is the Internet changing our morals and what we see as acceptable or are we simply using it to share where our minds already are more easily? The answer to these questions is actually quite profound and important because billions of people come in contact with these technologies every day, and the impacts are felt by us all.

What is not debatable is that people are constantly attracted to either the light or the dark in life. We are immersed in both every conscious moment of our day – they swirl around us. We choose whether we think about, look at, read, or participate in things that are healthy for us or unhealthy. All content we create and deliver through any media either will lift up the world or drag it down. Television always has given us a choice in that we have to turn it on and we have to choose a channel. It never forced us to watch specific content (at least in the U.S.). When we are young, we have little discretion as to what we digest, and that is a problem. When we provide easily available content to someone without them having an ability to discern higher and lower values we allow both into our lives, and then we must hope the darkness does not become us. As adults we need to understand that our higher purpose is to leave the world a better place, and that happens one person at a time. What you and I do online, what we watch on our screens, what we say to others over the wire either uplifts the world or tears it down.

I do not state this to be moralistic or judgmental about what television or online content producers are doing; I simply want to show the influence of technology on society. As I said, you can draw your own conclusions on the impact of television programming on society. The value here is to use television as an analogy for the impact tools like the Internet, online gaming, mobile devices and social technologies will have. If we stay ignorant to the impacts of higher and lower values on humanity, we will forever suffer the consequences. That is not to say that we should not have choices or that any one person - or one group - should have the power to decide what is light and what is dark. We have proven over the centuries that this form of control is not practical. There has never been a government, a human being or an organization that everyone would agree with when it comes to choosing what is healthy entertainment or content.

With that said we do have many governments, people and organizations that do what they can to filter morality. And since I am introducing the concept of filtering, let me mention that I will use this term generically in the book, and also as a term of art from the technology industry. By this I mean that we can "filter" data and information using technology tools in order to restrict the flow of anything we think people might not be interested in, or that we don't want a user to see. Obviously in the case of television, the "filtering" was done by standards groups and more through human observations than filtering with automated rules-based technology systems. Our courts said they could not define pornography, but they would know it when they see it. A computer knows exactly how it is defined because of the rules in its programming.

Technology Provides Mesmerizing Digital Entertainment

Life is not all about work for most of us and we try to fit in as many fun moments as possible. We have come a long way from when kids played with empty boxes in the front yard and adults played bridge around the table with their friends. One of the good things about having five decades under my belt is that I have experienced half a century of what it means to be entertained. When I reflect on where we have come from just at the beginning of my life to where we are today I am truly amazed at what has changed and what stays the same.

For example, music is a very consistent element in entertainment. We have had it for centuries and have created many new styles and instruments to produce sound. What has changed dramatically are the delivery systems. Centuries ago our families performed in our homes, we went to a concert in town or listened to the choir at church. At the beginning of the 20th century people enjoyed going to the bandstand and listening to the big band. Then there was a huge leap forward when we could record music and distribute it through the radio, then on LP records, then eight tracks, cassettes, DVDs, and finally with MP3s and other audio files. With each step we made music more convenient to acquire. The digitization of music also did something else wonderful; it provided a way for artists to connect to the world without the need of a delivery system controlled by agents, producers, radio stations, etc. Music lovers can now discover any kind of music from anywhere they happen to be standing, and at any time. And in many cases they can find it for free.

These changes shattered the music industry structure financially because the methods by which money was made changed dramatically. Artists now can promote and deliver their music directly to listeners and not even involve a label to do their marketing and distribution. The music itself even sounds different because of the all the new technology we can use to change the sound from origination to the listener. The fact that we have a device called an auto-tune says it all. The song may remain the same while how we deliver it is very different today.

Arguably the oldest form of entertainment is simply having conversations with other people. Again, the methods of achieving this have been changing rapidly. The telephone was really the first real-time device that allowed us to communicate in some way other than face-to-face and for decades it extended this form of entertainment. Today we have many other "real time" options including texting, social networking applications and video-based tools like Skype or FaceTime. Billions of hours are spent communicating with each other as a form of entertainment, and there has been a big shift from that being face-to-face or by phone to conversing virtually through pictures, videos and even email. Facebook could be said to be a massive form of entertainment where people share conversations of a one-way nature by surfing through hundreds of snippets in an hour.

Watching moving pictures has been a big part of our lives since we got the big screen movie theatre. We loved them when they were black and white and silent. We loved movies even more when they became full of color and sound. Soon after that we got the smaller screen in our homes. Recently we shifted to watching video on multiple sizes of screens and devices. There is a concept called the Theory of Three Screens which postulates that we will soon move to having a large screen in our homes for video content, a medium sized screen that goes with us when we work or need portable entertainment (tablet or laptop) and a small screen that goes with us everywhere (the mobile device). There will cease to be television per se. There will just be content and three screens. We will soon not differentiate between what is Internet content (like YouTube) and what was television. Millions of hours of video are now provided online and on demand, and most of it is not professionally produced. We all have the capability to be video content creators at this point, and what we create for each other is now providing a significant percentage of our entertainment. But really, cat videos? Seriously?

There is a fourth screen of course: the movie screen. I am sure that will always be with us because the communal viewing of new content on a large screen with robust sound systems is an experience that many of us really enjoy. In some cases, movies are now created and released directly for the small screen and skipping the step of even being restricted to movie theatres. Although this is happening more, I predict that we always will have some form of large communal screen with quality advantages over our smaller screens. Besides, we need a reason to go on a date or get out of the house from time to time. We will do this not because it is the only way to see the content; it will be because we want the experience of seeing the content in the communal format.

Technology and Games

I am sure games have been with us as long as we have been human. I am sure in the caves we used sticks and stones to play some type of competitive activity to pass the time. We are naturally competitive, which means we like to play games against others, we like to win, and we like the fun of at least trying. What it means to play games has changed dramatically just in my lifetime. When I was young, we played all kinds of made-up games in the yard with the neighbor kids. We played board games and we played card games. We played pick-up sports games for hours upon hours and imitated our professional sports heroes. Games were something physical, and most often played within six feet of our opponent.

Games are very different for young generations today. There are many screen-based options for entertaining oneself either against a computer opponent or against players anywhere else in the world. These games are designed to be addictive and have many levels of play so we can spend months conquering each one. They are often designed around a virtual world that is becoming more realistic every day. The goals of the games can be to construct a city, kill an enemy or to fit shapes together seamlessly. People play computer games for hours on end, some alone, some with friends in the room and some with strangers online (who they sometimes refer to as "friends"). It is often quoted today that the average 18-year-old child has played in excess of 10,000 hours of computer games. There is no question that what it means to "go play a game" today is very different from what it meant a few decades ago.

As I will point out many times in this book, new digital tools - in this case electronic games - have their plusses and minuses. Child obesity is rising in the U.S. and part of the reason is that many kids tend to get their entertainment from devices that lead them to sit in front of a screen for hours: television, their mobile devices and gaming systems. It is not unusual for a young person to come home from school and spend hours in a row sitting down and focused on a screen for homework, then for their modes of entertainment and communication with their friends. This is wildly efficient for accomplishing these three tasks, and yet devastating from a physical standpoint.

Other negative issues include the reality that today's online tools make it very easy for a young person to access any type of content that may be adult-oriented, be it designed for entertainment or not. As the digitized forms of entertainment get easier to access and more engaging, people get more distracted by them. Relationships can suffer through excessive connection with screens, and work or homework that needs to be done can get pushed aside as well.

Our underlying desire for fun has not changed though electronic delivery systems for our entertainment have changed the dynamics behind how we integrate moments of pleasure into our lives completely, along with the psychological impacts of the content we digest in the process. We now have access to entertainment at our fingertips every moment of every day, and from anywhere. I remember when I looked forward to Saturdays because that was the only day I could get up and watch cartoons. I remember listening to the radio station for hours just in the hopes they would play that one song I really liked. I remember reorganizing my day just to be sure I saw the one television show I did not want to miss. None of these are restrictions today.

I also remember when games to me were played outside with my friends and involved running, hiding and tagging, not shooting, killing or driving virtual vehicles around depraved cities.

I am not prudish about sex and violence. I have seen my share of movies and books with both and I'm old enough that I even have personal experience! But the ease with which kids can immerse themselves in games and content online that feature sex and violence is what concerns me. Once we allow a child to have a gaming unit or Web-enabled device, we lose much of the ability to filter what they are going to interact with or play. We can load software to monitor or control what young people see, but they can simply use a friend's device in order to get around this kind of restriction. Sexualized content and violence are aspects of life that are very real, and I am sure it is entertaining for young people to pretend to be part of these. I fail to see how letting a kid play *Grand Theft Auto* for hours on end provides any healthy aspects.

It used to be that all video entertainment came from a handful of producers who developed the movies and TV shows we watched. Today we can get on the Internet and find any type of video we want to watch, produced by any human with a mobile phone or professional-grade video camera and editing software. The array of music artists I can find in an instant is awesome. I can connect with anyone I want, anywhere in the world and talk to him or her with video through tools like Skype. No longer are we chained to intermediaries who decide what we watch, where we receive it or when we watch content.

As we look forward to the future, we will get more content from all corners of the world. We will get it with higher quality and extremely life-like virtualization. We will constantly seek for our entertainment to be as realistic as possible so that we can more easily escape our real world, if only for a few minutes. And this is where a huge new problem with our entertainment will exhibit itself: the addiction of escapism. There will be wonderfully uplifting content and there will be gut-wrenchingly evil content. It will reflect what attracts us and, as it gets more realistic, it will also have the potential to touch our souls in deeper ways. Our styles of fun always have impacted us, and this will continue to be true.

The Morphing of Our Minds

We have discussed how technology has changed our information delivery systems; now it's time to explain how technology actually can affect our physical brains.

Our minds dictate every single thing we do in our lives, so it is vital to understand how digital tool use changes our thinking. Recently, we have learned a lot about how the brain works and have developed the capability even to map visually what is happening inside our brains as they process information in real time. The resulting color-coded maps show the electronic activity that happens as we apply our minds to any specific task. This gives us a wonderful ability to understand how we process information based on the condition.

I was surprised when I first learned about the impact of technology on our brains in that I never realized it literally could change their physical shape. Further, the use of technology rewires the connections that allow us to process information and solve problems. Then again, I remember when Lumosity.com first came out with its brain training games. I spent hours testing them out, and what I learned was that they actually were effective in training my brain to process in improved ways. I remember having the vague thought back then that we probably were just scratching the surface as to how we might use technology to augment our brains in the future.

Neural research has been a watershed for researchers in understanding how any stimuli impacts how we process information and how technology affects our minds. Applying this research to the impacts of technology can give us an interesting picture as to how the brain of future humans will differ than our grandparents' brains.

How we memorize: There are numerous studies testing the impacts of technology on our ability to memorize facts or the type of memorization we do when we know we can lean on technology to help us with facts. What studies are showing pretty clearly is that our brains are adjusting to the new reality that we carry a memorization and fact-finding device at all times now; we call it a smartphone. The articles I have read on this subject paint the picture pretty clearly that we are slowly losing our skills of memorizing small facts (phone numbers, birthdays, addresses, etc.). This is due to the ability to store all of this on our devices. The more we trust our devices will store information for us dependably, the more we see no reason to strain our brains with memorizing thousands of small bits of data.

Nicholas Carr has been an outspoken critic of the impact of technology on humanity. His book, The Shallows, states generally that the Internet is having a detrimental effect on our minds' ability to memorize, which he supports with numerous scientific studies.

43

Betsy Sparrow, Assistant Professor of Psychology at Columbia University, on the other hand, sees this adaptation as positive. She says our minds are molding to the Internet just as they have in the past with technologies like the written word.

What is not up for debate in my mind is that mobile devices – our Outboard Brains – are changing how our natural brain processes and what it spends its time doing. We have only just begun to use smartphones that help augment what our minds do, and already we are addicted to the ways they provide instant information, store memory for us and solve problems. As we add heads-up displays, smart watches, and a host of new wearable devices we will continue to move the Internet closer to our bodies and our brains, so the impacts only will grow. What will be quite interesting is when a person's economic value becomes tied heavily to the technology with which they augment themselves.

How we process questions: When faced with the many questions that get posed in our lives, our brain is now operating differently in order to find answers. In earlier times we either had the answer readily in our brains we worked hard to deduce a possible answer by extrapolating facts we might know or we asked someone standing near us. Today we reach for our mobile device and get the answer in seconds. We are relying less on the people around us for answers, so now we divide the world of answers into:

1. I have it in my brain, or;

2. I can get anything I want from my outboard brain without having to strain my own.

This change actually is rewiring which parts of our brain light up when we are searching for answers to questions.

In an article written by Nicolas Carr titled *Is Google Making Us Stupid?*, Mr. Carr surmises that, by having a piece of technology to lean on, we are invariably going to be dumber because the technology will weaken our ability to dig deeply into content and process the information in ways that help us have better processing capabilities (my summary). As a rebuttal I would point to the Pew Research Center's *Internet & American Life Project* which asked its panel of more than 370 Internet experts for their reaction; 81% of them agreed with the proposition that "peoples' use of the Internet has enhanced human intelligence."

Also I must add my own observations. With the ease of being able to find any piece of information one chooses at our fingertips, we have a faster and better ability to get an answer to just about every kind of question. Therefore, many of us do a number of quick searches over the course of the day that we would not have done in years past. As with anything else, there will be some people who leverage it and those who waste this capability.

In the end, it makes about as much sense to surmise that Google will make us stupid as it would to say calculators made us all lose the capability to do math. The opposite actually has happened. The ease with which a calculator can do math has made more people proficient in doing more complicated equations that they would not even attempt if they had to do them by hand. As a tool the calculator enables just about every human being having the ability to do basic levels of math instead of limiting the number of people who are able to do this simply. Already we forget that, a century ago, there were people who specialized in doing "numbers," and everyone else simply lacked the ability to do what an eight-year-old can do today on a calculator.

Connecting with Each Other Exploded Overnight

Maybe not literally overnight, but social technologies leapt into our lives in what seemed like an instant. The ability to communicate with hundreds of people bi-directionally every day changes the way our brain processes in two important ways. The first is that we have the ability to "talk" to people every spare minute, so for many of us there is less "down time" for our brains. We fill these spare moments by connecting with a person or a group. Although it's a wonderful way to stay in touch with lots of people, it also robs our minds of time to process other kinds of information from our day. The second is that these connections with others may leach away time that had been spent reading or learning. Whereas in the past we might have been reading a book or watching an interesting show on TV, today that time sometimes is used to post status updates or read them from others. Granted, we can also get news from social sites, but my point is that we are not always aware of the raw material we feed our brains on a daily basis. Social tools are wonderful for connecting us, and they also fill time that was invested in other, sometimes necessary, ways.

For many decades we have had a limited set of options as to how we connect with other humans. The primary way has been face-to-face, with another being the telephone more recently, and of course we have used the written word, generally on paper, for quite some time. These physical methods of communicating can be rich in context, and at the same time, can be inefficient. Before the 90s, we would not have understood why, because our frame of reference had not yet changed radically as it has today.

In 1992 British anthropologist Robin Dunbar published an article stating that, in primates, the ratio of the size of the neocortex to that of the rest of the brain consistently increases with growth in the number of entities in a social group. This is true for primates at least. As an example, the Tamarind monkey has a cortex-to-brain ratio of about 2-to-33 and an average social group size of about five members. On the other hand, a Macaque monkey has a brain size ratio of around 3.8 with a very large average group size of about 40 members. Based on this work Dunbar developed his "social brain hypothesis." The relative size of the neo-cortex rose as social groups became larger in order to maintain the complex set of relationships necessary for stable co-existence. Most famously, Dunbar suggested that given the human brain ratio humans have, we can expect a social group size of around 150 people. This is about the size of what Dunbar called a "clan."

In plain English, Dunbar states that the size of our brains literally can change based on the level of social networking use. Actually this is not surprising because our brain changes sizes whenever we focus a lot of our attention on any specific thing, like juggling, for example. This is called neuroplasticity. So don't be shocked that our brains are changing in relationship to technology use; just be aware of our capacity to evolve biologically to adapt to technology.

The following is a very real statement: This was your brain in the 90's, and now this is your brain on Facebook. The heavy use of social technologies literally is changing the shape or our brains as we process information differently. That is fascinating to me.

As you may have noticed, I have made it a point to remember the first time I used a new piece of technology. When it comes to social technologies, I remember when I first visited an AOL chat room. Although we don't think of AOL as being one of the early social networking providers, they were. There were even services before theirs (e.g. The Well) that leveraged an online capability to connect strangers and friends online. Both of these systems were replaced with online communities like MySpace, which of course I remember setting

up one day. Then later it was Facebook that took over the leadership of social community building. I was walking down the hall of our house toward the back bedroom and my wife was passing me going the other way. I asked her, "have you tried the new site Facebook yet?" She said, "I don't have time for things like that." About a year later, she signed on for the first time and discovered that a high percentage of the people from her small Oklahoma high school were already on Facebook posting about their lives. She was hooked immediately by this. Later she discovered she could stay up to speed on our daughter and other family members the same way. Being an extrovert she has, to this day, made social tools a big part of her day. Whereas in the past she could only stay in touch with five or ten people a day, social tools have given her the ability to stay current with 50 to 100. She now tells me that this capability keeps her from being lonely when I am gone. If you step back a bit, that is a profound capability to be provided by a bit of technology that fits in one's hand.

How We Connect with the Universal Mind

I have read a number of science fiction books that play with the ability simply to download information directly into our brains just like a hard drive. In one particular story the protagonist would download new languages into his brain so he could immediately talk with any native speaker. I had the impression even back then that it might be possible to do that at some point. Since then I have watched our scientists develop brain-to-computer connections that allow our brains to be wired directly to computers. This type of interface is called a BCI (Brain-Computer Interface) and we will talk about it a number of times in this book. The reason for this is that the BCI will represent a huge new wave of change in the integration of man and machine, and it is certainly going to go mainstream in the not-too-distant future.

At the point that it becomes common to augment our brain by providing a direct connection between it and the Internet, we will have extended our minds to what I call the Universal Mind. Instead of just having access to the limited information our brain holds at any one moment, we will be able to tap instantly into the collective knowledge of everyone around the world. Lest you think this is too far off or too crazy to consider, we actually have this ability today through Internet search capability. When we do a Google or Bing search, we are tapping into the collective knowledge base for answers. The only difference is that we have to use our fingers or voice to do the search and the response is delivered on a screen or verbally. With a fully developed BCI, we will be able just to think about a question and have the answer pushed back to us instantly. The impact of that capability will rewire our brains yet again because we will be less individuals with our own sequestered thoughts and more part of the Universal Mind where we share our thoughts and questions routinely.

It is not a stretch to see that as we can connect to the electrical impulses in our brains today with our early attempts to read what our brains are processing, we might one day be able to reverse the process and put information back into our wetware. Regardless of the direction of the information flow, the fact that we now can connect our brains to a machine is staggering in its implications and we need to really consider the ultimate impacts to us of this integration.

How Technology Affects Spirituality

There is no question that technology has played a huge role in improving our quality of life and will continue to do so. We will be more productive, have more flexibility, have more options for entertainment, and all of this will get easier and easier to achieve. Humans are on a never-ending quest to make life easier, and to live lives with more positive experiences and fewer negative experiences. Technology is an increasingly powerful tool to help us on this quest. When it comes to happiness we are on our own to make that happen because it seems no amount of technology integration into our lives will do it for us.

Technology is changing how we memorize, how we process information and how we get answers to questions. And still it is in a very early stage of full development. Processing power is growing and our ability to gather, store, and analyze data is expanding rapidly. As we talk more about the future we must examine how these major functions in our lives are going to change our world.

We are falling down a slippery slope (or making a great climb) toward the integration of our brains with technology, and that is going to change our species more than any other tool we have seen in human history. There is nothing with more potential impact to alter our brains and how we are able to connect our minds to each other.

From a spiritual standpoint, the fact that technology is rewiring our brains is an intriguing situation. We normally think about anything spiritual as standing in opposition to technology and its connections. For many there is a visceral negative reaction to the news that technology is rewiring our brain, and to it being connected directly to our brain. We think of being spiritual as meditation, deep contemplative thinking about God and the peacefulness of being out in nature with nothing to distract us from communing with the Spirit. What if our ability to more easily connect with information (correct information) and with others is a manifestation of holiness? What if our minds are headed towards the Universal Mind (Oneness) and that is the direction of holiness?

Is it possible that technology will help connect us in a much deeper way so that we are not such separate islands of thoughts and emotions? With the aid of technology will our brains mutate toward having the ability to share thoughts instantly or ask questions with others? Some would argue we are already there with the advent of social sites and mobile tools. If we are willing to step back and look at the Internet as simply a powerful connection framework, it becomes clearer that as it evolves, and as our brains evolve with it, we very well may be headed towards a computer assisted Oneness as a Universal Mind.

Our Digital Addictions

Technology not only has changed our physical brains; it has affected our patterns of addiction. I have been reading an endless stream of articles about the dangers of technology addiction. Let's define what the word "addiction" means so we can have a coherent discussion on this topic. An addiction is a compulsive, unnatural and/or unhealthy desire for something. The word "addiction" is generally used to describe a negative attachment, so when we use the word, we are inferring an unhealthy state of being. People can be addicted to anything: exercise, food, sex, drugs, alcohol, gambling or spending money. Additionally, we can be addicted to things some of us do not normally recognize such as being nice, keeping our mouths shut when we need to speak, controlling other people, rescuing other people, venting anger or cutting ourselves. Simply, an addiction is where we put an inordinate amount of our personal investment into one thing, whether that thing is considered a healthy or unhealthy activity.

When personal computers came into being at the turn of the 80s we unleashed the ability of an individual to control computing power while sitting in their living room. This was a huge step in technology proliferation because people who were willing to invest the time to learn how these devices worked were able to make the PC dance to their tune. For some it was a means for writing software that could solve specific problems, and that could be sold or shared with others. For others it quickly became a second screen (TV being the first) they could be entertained with. Whenever we develop new devices, we quickly figure out how to use them for four things: finding information, work, play, and communication.

The addiction to technology ramped up because, for the first time, a person could tailor their PC to deliver whatever filled their specific needs or desires. It started slowly at first because few people really knew how to write software, while as interesting software applications were developed and more people bought PCs or used them at work, the attraction grew. It would be the Web that really would lock in the addictive properties of the PC, and that came along in the mid-90s. It is important to note that technology at this point was not creating any new types of addiction; it simply was delivering the addictive content that people had previously gotten on paper or in person. For example, the most well known addictions we fed early on were the cravings for porn and shopping. These activities were nothing new in the world, however the easy availability just exacerbated the addiction.

The more prevalent addiction today is to our smartphones. It started with young people who quickly developed a very emotional attachment to their devices. They often sleep with them, eat while looking at them, and panic if they cannot find their phones at any instant. It did not take long for their parents to follow the trend as we became addicted to them to stay in touch with each other and to get our daily tasks done in easier ways. Even grandparents today are becoming quite attached to their mobile devices for the safety aspects and the ability to communicate easily with friends and family.

For adults, mobility connects us to our kids, to our careers, to information, to shopping, and to most of our friends and family. When I give speeches and begin to talk about our addiction to technology, Often I will grab a mobile device that is sitting in front of some businessperson on the table. I will walk away with it and put it in my pocket. At the end of that section of the talk I will look back at the person whose mobile I borrowed and they will be visibly agitated. I will ask them how it felt when I walked away with their mobile device and they will always say they were a bit anxious.

Is it really the device people are addicted to or is the connection the device gives them? The addiction really has nothing to do with the device; it never did. It is a combination of the convenience of services the device provides, the instant connection to others and the security it provides in times of need. As an analogy, it is not drugs people are really addicted to; it is the feeling the drugs give them. If we want to help heal people we must fill the hole inside of the person that drives them to take the drugs in the first place. It is the same with food. It is not the addiction to the food that is the problem; this is just a symptom for most people. The hole inside that they are trying to fill is the cause of the addiction to eating. This has always been the case in my life because I have long looked at food as a reward or as the fun in my day. This drove me to eat things that made me happy, not things that my body needed to be healthy.

This distinction is important with digital mobility because we then can see that it is not the physical device we are addicted to: it is the convenience and connection the mobile device represents. Think about most teenage girls that are heavy mobile device users. They don't use the device to make phone calls as much as they text, Facebook, Instagram, Snapchat, search for information, watch videos or listen to music. So the device provides a connection to their friends, to information they want and the entertainment they enjoy. Because they can have this connection all the time, it creates an environment where they are always connected to the things they love.

At that level the smartphone provides a magical capability that no other generation has ever had – not even close. Their mobile device is the conduit to friendship, knowledge, and entertainment, all on demand. Of course young people are addicted to it. Friendship, knowledge and entertainment are just about the three most important things in a human being's life.

I think it is critical that we all really come to understand what is healthy and unhealthy about this addiction to the connections our new devices provide. It is not a right or wrong decision as much as there is a continuum of healthy to unhealthy behaviors within the shell of our desire to have all this power at our beck and call. Let me illustrate with a couple of examples. On the healthy side, the ability to connect with your friends more often is a great and wonderful thing. Connecting to them by texting while you are driving is not. On the good side, the ability to find any piece of information in a nanosecond is a wonderful new capability for humanity. The ability to instantly read twenty articles online about how to poison your spouse is not. The ability to see the Top Secret Drum Corp on YouTube from any place you happen to be standing is a great capability that we have never had. The ability to video someone without their permission and upload it is an invasion of privacy.

The intriguing thing about us being addicted to our new means of connection is that God is all about connection. God is love, and love is about connecting with God and with other people. For this reason, it could be said that God is the Internet in some ways. The Internet binds us and mobile devices are our physical connection to the Web of people, information, and content. Ergo, we might be wrong to indict ourselves over the "addiction" to being connected. Surely we can agree that misusing our new connection power in ways that would harm other people is bad. Consequently, we also have to acknowledge that we were created to connect with others and to God and in this way, we were also created to be "addicted" to the connection that technology is giving us.

New Relationship Dynamics

Because our new capabilities give us powerful ways to connect with people and things around us we will have a growing capacity to choose which relationships we invest our time in. We only have a finite amount of time to invest in the things that are important to us in life, and nothing is more important that what or who we spend our time currency with. Up to this point for most of us the word "relationship" referred to people connecting with people. That is about to change drastically....

There is a pretty sound argument to be made that we already have introduced a new relationship into our lives that is soaking up a large percentage of our time: that is our connection to the Internet. Before the Web, we could spend time with people, pets, on musical instruments, etc. These were, and still are in some cases, examples of where large amounts of our daily minutes would be invested. If we could measure the migration of minutes from these activities to time spent online, I am sure we would be shocked at the transition. Although it is true that when we are using our devices to "talk" to other people, we are in a way relating to another human, it is not the same as a face-to-face conversation. So in a manner of speaking many people have a very strong relationship with the Web and all that it provides. As I mentioned we only have a finite amount of time in each day so the time spent relating to something through a screen means time not spent relating to a person, a pet or on an instrument.

I am not trying to address all changes to relationship dynamics in this section: just the concept of what we are relating to and how that might change. Going a step further, I want to look at the entities that are even in relationships. For example, there will be people in the future who form very emotional connections to a piece of technology that interacts in very human-like ways (e.g. an AI, android or robot). And there will be relationships between physical devices that become much more sophisticated.

Human-to-Machine relationships: The movie *Her* was an eye-opening example of a person getting lost in a relationship with a machine. It tells the tale of a man who becomes emotionally attached to an operating system that communicates with him with an attractive female voice. The voice is only the beginning because as the operating system gets to know him better it begins to fulfill a role in his life of a great friend. As the movie progresses he becomes more attached to the disembodied voice until an awkward scene where the operating system hires a call girl to come to his apartment to be a stand-in for the program being there in person. The main character cannot connect with the flesh-and-blood girl because he knows it is not the same as what he is relating to on the computer. As a human being there is a sense of loyalty that he cannot get past, even if the call girl is pretending to be the voice from the computer. You can watch the movie yourself if you want to see the ending.

In a way, we already are taking the first steps in building relationships with inanimate objects when we build a strong attachment to our mobile devices. Even though they are not able to relate to us in human ways (other than taking voice commands and questions, and talking back to us), they are playing a really critical role in our day. Our vehicles are much the same. As we put more human-like interfaces in them, and as they become driverless and just take us where we want to go as if they are an independent friend, we will see them as being more "alive." Already today many people name their vehicles and associate a personality with them. We do this because we are so connected to our vehicles, the critical roles they play in our lives and the amount of time we spend together. This is a slippery slope because, once we become dependent on machines and give them more human interfaces, naturally we will treat them more as friends than machines.

When I describe this concept of building personal connections with machines we could be talking about a house, a car, a mobile device, a robot or any kind of computer-driven piece of equipment that relates to us in a human way. It is not the device we would get emotionally connected to, but the operating system or application that is powering the device. Technically it would be the human-like software construct that helps the device talk to us. If you are getting a little uncomfortable with this line of thinking, let's look at where we are already with assigning human qualities to things that are not human.

A device with the right kind of software can talk to a person, help them find information, anticipate their needs, and entertain them. In some ways, the device is a lot smarter than an animal. So is it really that strange to understand that people increasingly will become attached to their machines? There is one thing animals provide that not all machines will be good at: physical touch. Animals will cuddle with their owners to fulfill the need of a human to touch and be touched. This takes us to a further evolution of human and machine relationships – between a simulated mechanical human and the real thing.

Why would you guess people are so intrigued with robots? Why do we portray them so often with human-like qualities? It is pretty obvious that the vision we have as a species is to create machines that are a very close approximation of a human. The big difference - and one of the huge problems - is that we want complete control over them. We already fear the possibility that we will build human-like robots, then they will confound our ability to control them and go rogue. In other words, already we seem to understand that one day we will have robots that are very human-like, and already we are on to fearing the next step. An older movie that played this script perfectly was *The Stepford Wives* which came out in 1975. Over forty years ago we were projecting what things might be like if someone were to create a robot that was hardly distinguishable from a human.

Many people seem to love the thought of having a machine that looks like a human, works like a human and takes orders like a slave. This would allow us to have complete control over a machine - which we equate with being a person - in order to get exactly what we want out of life. This includes having "someone" who can get our work done for us, like the Roomba robot vacuum cleaner we already have in the market. We just want the Roomba to be taller, have speech capabilities and a great looking body. This humanoid could keep us company, help us do our work and even fulfill our every physical desire as programmed. Privately, many people love the thought of having this kind of "mechanical slave" that they believe would help them have a better life.

Don't be naïve and think this will exist only in science fiction. The relationship between human and machine is already strengthening and as we continue to build more realistic robots and add humanoid qualities to our machines, we will continue to build emotional connections with - and dependence on - things that are mechanical. I do believe that the emotional relationship shown in the movie *Her* between a person and a machine will happen in the not-too-distant future.

Machine-to-Machine relationships: You may be thinking that this cannot be a real relationship because only humans have real relationships. That, of course, would discount the relationships that animals have with each other. We have crossed the bridge into the M2M (machine-to-machine) world already because devices are being built to "talk" more easily to each other through data exchange, and over the

Internet. For example, our mobile devices use their GPS capability to know where they are and automatically will allow software applications to use our location to push alerts to us. Smarter devices now are gaining the ability to connect with each other to compare data and their current state to provide humans with more safety or convenience. The "Internet of Things" era actually is based on devices being able to "talk" to each other and to be controlled by humans on a whim.

As the devices, machines and technology get more independent and autonomous they will connect and communicate with each other in very different ways than today. We will live in a world where, as devices are purchased or come in physical contact with each other, they will recognize the other devices around them or in their performance realm and they will connect and share information and instructions. It will seem to us that they are alive and intelligent because they will be programmed to know how to connect and "talk" to other devices. This will change how we view, and how we feel, about the devices, machines and technology around us. For example, I believe one day our cars will know our homes, they will have a relationship of sorts where they discuss various things about what each of them knows about our activities. When our car gets home it will know it is home and our home will know our car is in the garage. Our car and our home will have a discussion when they are physically close to each other.

They actually will start sharing information when our car gets in range of the house. The car will tell the house who is in the car in case there is a guest coming to the house, and maybe even the state of mind of the driver and passenger. The house will start adjusting aspects of our homes based on what the car has told it in advance of arriving. The car also will report any mechanical problems it might have so our smart home can take some early action to help us arrange to have our vehicle serviced. Based on the music I am playing in the car, my home might automatically transfer the channel to my home so it continues the music I was listening to. If I am on a call, the call may automatically transfer to the hands-free audio in the house so I can walk from car to home and keep talking.

Because our home network will talk to all the other appliances and systems in the house, many things can be turned on or started as our car gets within range. The temperature can be set where we like it. A meal could be started or a bath could be filled for us. The list goes on and on. The M2M conversation my car and my home will have will help me to be safer, provide convenience and will save me loads of time. We will become very addicted to all of the machines around us having relationships with each other to benefit us.

Our mobile devices will know us and will customize to all the specific rules and services we want them to provide for us. They also will build a relationship with all the other devices in our lives. They will know our car, our home, our appliances, our office, our family's mobile devices and so on. And because they have a relationship with all of these they will change how we live because it will become the controller that enriches all of the machines, robots and technology around us. It truly will feel like the devices around us have come alive and have gotten to know each other in order to help us. Already we have taken steps in this direction when we load mobile apps on our smartphones that allow us to control lights, alarms, door locks and other devices in our homes.

As I mentioned earlier, the machine-to-machine relationship is one that we don't think much about in terms of technology changing our lives, yet this may be the relationship that changes our world more than any other.

We are certainly in a time of transition with relationships, including who and what they are with, how we find them, how we build and maintain them and even how we end them. We think of technology as a tool that helps us get things done more easily. We rarely consider how it impacts all of the relationships around us. We let it distract us from the people who are in the room with us. Often it dictates who we talk to each day and how long we spend communicating with them. The actual people we have relationships with is highly impacted by whom we can connect with through technology.

Soon we might struggle with a situation where we are getting more from a relationship with an operating system than from the people in our lives, and that may unleash a whole new spectrum of psychological issues. We will also move into a world where machines, devices and robots all have relationships with each other that we control at some level but do not control completely. In the middle of that world where devices around us come alive, so to speak, we will be relating to this net of technology to improve our quality of life. As much as this might sound like science fiction to you at the moment, we are very close to the front edge of this now.

That makes it very important for us to take a step back and be aware of all the ways technology is impacting our relationships in the world. This is an area of technological change where we must make good decisions about how we react to the implications so we improve our relationships in the world instead of cripple them. Our relationships with each other are what drive the vast majority of what is good in our lives; if we cripple our ability to have close and meaningful relationships we will have crippled much of what makes us human.

SUMMARY:
Technology, Outcomes, and Choices

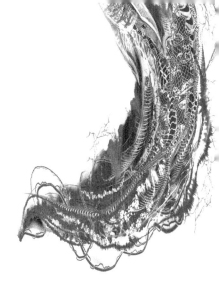

It truly is hard to comprehend the amount of change that technology (going back to the 50s) has brought to humanity. Couple that with the fact that the pace of change is still speeding up. There are few places in our lives where technology is not altering our behavior. With each new digital invention we have a choice about how we use the tool provided. Any tool can be used for progress or to disrupt essential life processes. Now is the time for us to be awake to the impacts of technology and to make conscious choices about how we use it.

Over the past five decades we have embraced technology as an almost magical tool that can improve our lives. However, there are negative aspects of, or uses for, any tool. What we build also can destroy, and what we connect with also can disconnect us from what is more important for our wellbeing. The good news is that we always have had choices about how we integrate technology into our lives. The bad news is that too many of us have been unconscious to the impacts of the choices we are making today. The danger only escalates with our future digital tools.

The relationships we have in life inform the vast majority of what is important to us. Humans are built to relate to one another at a number of levels including physical, emotional, intellectual, spiritual and through shared experiences. Technology has a growing impact on our ability to find, grow and manage relationships because it improves our abilities to do each of these. It provides a way to scale the volume of relationships we manage and tools to improve our ability to communicate with people we are relating to. This new ability to leverage technology in our relationships is opening up huge new capabilities to deepen and maintain relationships across great distances and to restart relationships that once might have been lost to time. These all are wonderful improvements in our quality of life.

At the same time, technology can pull us away from developing healthy relationships without really being conscious of the trade-offs we are making. In an attempt to stay involved with many online-supported relationships, a person can ignore the people closest to them, both physically and emotionally. The easy access to porn or unhealthy online relationships can devastate marriages. An overreliance on digital connections instead of face-to-face relating can lower the quality of what otherwise could be a deep relationship with someone close to us. Maybe the most dangerous impact of social technologies can be that they lead us into trading quality for quantity. People can fall into investing two or three hours a day keeping up and communicating with people in online communities instead of investing that time in their closest family and friends. It may not seem like much of a trade-off, yet all of those hours add up. Over time a person can find themselves well informed about the lives of 200 friends online, but without a single deep and caring relationship in their face-to-face life.

My mother once shared this with me:

> *Blame of the external in any way will defeat us. Freedom always involves looking in, not out: changing our own littleness into our own magnificence.*

Humalogy Viewpoint

Technology exists in the realm of the physical and it can make our lives more prosperous, easier, more connected to others and more entertaining. Used in a healthy way, it can provide ways to feed our souls. It is time for all of us to understand the concept of the Humalogical integration of people and technology tools in order to get processes done, and that most of us will integrate with technology at pretty deep levels from this point forward. While this is happening we must balance this integration by strengthening our spiritual maturity so that we do not slip into being androids with empty souls. There is huge value in what is uniquely human (e.g. love, joy, and peace: all qualities of the soul) and we need to hold onto these as what ultimately will fulfill us.

My concern today is that we are grasping blindly at technology as a catalyst to improve our lives and organizations without understanding the very real negative impacts. Please understand that I am a technologist by career and I love what these tools do for the world. I want you to be able to be conscious of the integration that is going on now and be able to judge how to balance what is uniquely human with the tools that provide magical new capabilities. There is no doubt that we as a species are moving from the H side towards the T side in how we operate each day. What none of us can know for sure is just how far we will march toward integrating our bodies and our very beings with technology over the next 500 years.

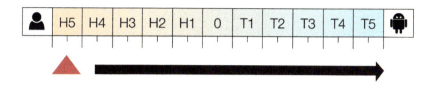

Seminal Questions

Will we become more aware of the potential impacts of future technologies on the human race now that we have television and the Internet as examples?

As we integrate our minds with an outboard brain (mobile devices), will the resulting combination be better or worse for the human race?

Ultimately will we create a Universal Mind that transforms humanity to its higher function?

Will too many of us allow ourselves to be more distracted by technology than investing our time wisely to deepen relationships?

Will technology drive us to have more relationships by quantity and, at the same time, fewer deep and meaningful relationships?

How dangerous might it be for a person to build a relationship with artificial intelligence built into a machine? Will this simply help lonely people have more connection with something that resembles a human at least, or will it drive a dangerous psychosis in humanity as we choose machine-driven relationships instead of connecting with each other?

How will life be different when devices around us come alive because of the Internet of Things? As our quality of life goes up will our abilities to be self-sufficient shrink at the same time?

CHAPTER THREE:

Digital Connections, Relationships, and the Changing of Tapestry of Life

There are three kinds of death in this world. There's heart death, there's brain death, and there's being off the network.

—Guy Almes

One of the most powerful aspects of technology is its ability to connect various areas of our world that have never had connections before, and to do it instantly and very inexpensively. Connection is a good thing at a spiritual level when it provides a way for us to flow love and support between us. Connection is a good thing when it flows knowledge and help between us. We were designed to be connected to each other and nearly everyone has an innate desire to connect with other people in a deep way.

The connections in our lives are a lot like our lungs breathing or our hearts beating. We take them for granted until we lose them and then we face serious consequences. We walk through our day connecting with nature, parents, children, friends, co-workers, bosses, strangers, and information sources. As long as these connections are healthy and fulfilling we don't think about them all that much; they simply add texture to our day. However, when a parent or mate dies, a child leaves home, a friend betrays us, a co-worker fails us, a boss fires us, a stranger harms us or technology breaks down, connections are broken and life never is the same.

Our lives are tapestries of connections between people, groups and organizations. We are also now deeply connected with our new digital devices and they both facilitate connections and provide a very general connection for us to a vast amount of information in the world. This tapestry is changing in its complexity, size and feel as the possibilities for connections grows exponentially. Each of our individual tapestries is not perfect in love and harmony, nor will it ever be complete while we are alive. We have holes and threads hanging out where we have made mistakes in life or where connections have been broken or lost. The Internet is a powerful loom that is helping us to weave this new tapestry with all of its benefits and dangers. We now have a Humalogy-based tapestry where we get to choose how we balance digital connections with physical ones. Whenever we make a poor choice about how much "T" vs. "H" we choose, there is the potential to compromise our connections.

We define ourselves by the connections we make. Because these connections allow something to flow between us we are fulfilled or depleted by that which we allow to flow or that which is restricted. Isolation is so intolerable to human beings that prisons use segregating prisoners as the harshest form of punishment. Being dis-connected is normally a negative state whether we are talking about the Internet, a relationship with a family member or friend or our own soul. The more connected we are to each other, the more of a sense of well-being we have, so it is logical that the more technology can act as a catalyst to connect people, the more positive it is.

Digital tools have offered us a broad new list of choices of content to connect to and the channels we use to connect to it. The blessing is that we have many new and inexpensive means of connections and that creates a new volume of connections in our life. The curse may be that we are giving up quality for quantity. In other words, we have hundreds of "friends" online, but no one with whom we really have a meaningful conversation at the heart level; connections with people who would be there physically to help us if something went wrong in our lives.

Importantly, technology has the capability of allowing us to connect at a head level to hundreds of people a day. It also has the potential to disconnect us from others at the heart level when we focus on the quantity of connections instead of the quality of connections.

Let's take a deep dive into what connections really mean to us, where they may be going, and why we need them in life. At a soul level, I believe we were designed to connect, and this is why people wither and die when imprisoned in solitary confinement – in a very real sense, a fate worse than death. We were designed to flow love to each other and, when done genuinely, both the giver and the receiver are enriched. It is clear that we progress as humans much better when we connect in loving ways. We are all connected to Earth with gravity and are seldom aware of it, yet no one has fallen off the planet. Our souls also are held with a kind of gravity called love, and we never fall out of that either. Maturing as a human being means becoming more aware of that soul connection; at some level we all know it is there. This is the archetype for all connection.

The Internet provides an invisible connection that allows all humanity to flow many things to each other in ways that can enhance both the giver and receiver. In this way, the Internet is very much a God-given gift. Perhaps this is what Pope Francis had in mind when he stated as much.

In order to understand how technology impacts us, one must really see how technology connects elements that never have been connected as easily before. It is through these new connections, or the shattering of our current methods of connection, that we will benefit, or struggle. For example, a smartphone connects us to information, to each other and to the things around us. It does all of this at a scale and speed we have never seen before in humankind. The power contained in the smartphone to help us, entertain us and educate us is unprecedented. At the same time, the smartphone distracts us from driving, from learning and from each other. Any one of these three can have serious negative consequences.

The Internet simply is the most fantastic tool we ever have assembled for providing connections between human beings, companies, and devices. This could be considered on a purely physical level, or more importantly we could extend this amazing capability to the spiritual level. As the Internet connects each of us into a Universal Mind, we can more easily share our gifts. Maybe it would be possible to stop right here and admit that God did invent the Internet because, when Jesus was questioned by the Pharisees in the New Testament about what the greatest commandments were, he was very clear that there were two: Love (connect with) your God with all your heart, and love (connect with) with your fellow human being/neighbor as yourself. Nothing has connected us to our fellow humans like the Internet!

Our world turns on what connections and separations. When two people have a connection, progress can be made in their lives. When they are disconnected, pain blossoms. When two devices are connected, information flows and the devices can better be controlled through those connections. When a device is disconnected, it cannot be controlled virtually and data normally will not flow to or from it.

Think of the difference between an automated sprinkler system that is analog and cannot do anything more than be programmed in the garage vs. a "smart" digital sprinkler system that can connect to a smartphone and tell its owner that it is raining and does not need to run today, or that it is hot and needs to run twice. One is disconnected from the homeowner and one is connected.

As a way to create context for you, I want to divide into five separate eras how I see technology connecting parts of our lives that have not been connected as powerfully before. These do not have hard beginnings and endings and should be viewed more as stages of growth that overlay and blend. By naming them I am just pointing out distinct types of connections that we already have or that will flourish one day in the future:

Web 1.0 – **Connected Organization:** In the first iteration of the public Internet we were exposed to the Web as a concept, and the first generally-used exhibit of this was a Website. Websites could be built for a low amount of money, and they were accessible by anyone in the world with an Internet connection. Although some individuals built Websites to advertise themselves personally, the vast majority of sites were dedicated to organizations. What quickly became obvious was that entities now had a powerful tool to connect with their constituents in ways they never had before, and at a scale and cost that was very different from traditional means of connecting. Instead of face-to-face, telephone, direct mail or media-based advertising, organizations now had a method to "talk" to people instantly anywhere in the world.

This ability changed entire supply chains and transaction methods because the supplier of a product or service now could sell to anyone in the world for a much lower cost. When that happened, we had airlines selling tickets directly to travelers. This had a devastating impact for travel agencies, as two-thirds of them closed down over time. At the other end of the spectrum catalog companies then could forego spending a fortune on paper and postage instead of posting their wares online using much less expensive techniques to drive buyers to their Website. Their new catalog Websites were, by nature, international, enabling them to sell anywhere in the world overnight.

There were winners and losers in the economy when the Web first showed us its power to connect entities that had not been connected in this way before, and this was a prelude to even more dramatic changes to come. Geography began to be much less a factor in the economy. The size of an organization became less of an advantage. And the marketplace, which once was limited by drive time, began to unshackle itself and form new connections directly between consumers and suppliers.

Web 2.0 – Connected People: The next Web era picked up traction around the turn of the century and was defined by social technologies, mobile devices and cloud computing. This mixture allowed people around the world to talk to each other at a very low cost, instantly, and at massive scale.

Just like Web 1.0, the new ability to connect people in ways that could not easily be controlled by governments or through traditional media outlets dramatically changed humanity and the economy. Now it was possible for any one person in the world to talk to everyone else with an Internet connection instantly and for free. This created the concept of Citizen Journalism where any person now could deliver the news or share an opinion. Groups of people now could form around any topic of interest and geography did not matter. These groups could be Volkswagen lovers or revolutionaries planning a protest.

In the world of media, Web-connected people completely changed the game. For print media companies, the problem became the decreasing desire to get information printed on paper. Besides killing trees (if it was not recycled) print media also delivers information slower than it can be received online. And the costs for delivering information through print are much higher than online and someone has to pay those costs. Consequently, we have a young generation growing up who rarely chooses to read words on paper if given a choice. This might seem a bit ironic if you are reading this in paper book form right now! The migration towards delivering content in digital forms is forcing new business models on many print media companies, and chasing a few into extinction.

Radio and television are faring better, but struggling with lower advertising growth rates as sponsors shift their marketing dollars to digital marketing schemes. The advantage digital marketing has over the print business is that digital content can be delivered very quickly and without regard to supply costs. A digital piece of content is replicable billions of times without much cost. In the broadcasting realm audio and video are just as compelling on a mobile device or computer as they are on a TV or radio. The difference is that the online delivery of video and audio is more targeted to markets and self served when and where people want it.

The major point to take away from this is that, once again, a new era of the Web is rewriting the rules for the economy and society in general.

A good example of the connected-people era is what happens now on your birthday when you are connected to social sites. My birthday was a couple of days ago and I received about 75 well wishes through various sites like LinkedIn, Facebook, over email, etc. I received six paper cards from those close to me and one phone call. The math might be different for your birthdays but I suspect the ratio might be close to mine. Let's set aside for a second the quality of the online connections since some of them said nothing more than "Happy Birthday" in the text line. Think instead about the positives of having many more people connect with me than in the past. Of the seventy-five digital connections, there were a handful from people we have not talked to in a long time who communicated in a deeper way, and we responded in some detail. This started a conversation that we were happy to have with them. In some ways we dread all the connections every year because we feel like we need to respond to each one separately. In other ways we appreciate that on top of the seventy-five people who reached out, there were hundreds more that at least saw it was our birthday and for a few moments, connected us to them in their minds. Hopefully for some of them, they were connecting also in their hearts. I love the cards and phone calls from people because they have the most meaning, and that does not diminish my love for the number of people who reach out virtually now to acknowledge this day with me.

Web 3.0 – Connected Devices: The third era of the Web will be defined by intelligent devices with an ability to connect to each other, and to our mobile devices so we have better control over them. These smart devices will broadcast data at a rate that is hard to fathom because we have never had everyday, inexpensive devices that could create, gather, and broadcast data in this way.

We are already deep into this era because we have begun to empower many of the devices we use every day to be connected, and are adding intelligence to them at a high rate. We already have smart homes, smart cars, digital sprinkler systems that talk to our mobile devices, and Web-enabled appliances. We are adding intelligence quickly to every device we possibly can so that they are driven by software based rules that automate their behavior and allow them to talk to us about what they learn. We even have a Bluetooth toothbrush that can give us information about our teeth and the brushing experience.

The changes that will be brought to the world when we integrate our lives with devices that all can connect with each other, and with us, will be at least as dramatic as the previous two eras:

> *The number of devices connected to IP networks will be nearly three times as high as the global population by 2017.*

> Source: "The Zettabyte Era - Trends and Analysis."
> *Cisco VNI: Forecast and Methodology*

Web 4.0 – **Connected Information Streams (Ambient Intelligence)**: This wonderful new era will be the outcome of our current work in developing the areas of Big Data and Business Intelligence. Picture this era as a huge number of rivers of information flowing all over the world for anyone to tap into at will. Picture yourself looking through a data listing site with thousands of free sources of information and being able to click on any source of data you want – soccer scores for the Premier League, governmental performance data, running costs for your vehicle, weather trends, or the sales of products at Starbucks, and pouring that data into your information management system. Then add massive amounts of data coming to you from all the devices you have purchased, your home, your appliances, the wearables on your body, and those of your kids. What we are seeing today is just a trickle compared to what we will have in the near future.

With all of these sources of data we will not only be able to see forensic information, or real-time status, we will also have predictive analysis capabilities overlaying just about all forms of data we are connected with. In order to fully utilize these massive rivers of information, we will need a software-based tool to help us capture, filter, manipulate, and analyze all of the information we connect with. The combination of data flowing on a scale we cannot imagine today, with powerful information management tools, will define new capabilities that will launch Web 4.0.

As we get better with gathering and harvesting data, we will also add to that the ability to flow our customized data streams to whom we want, when we want, and in real time. Although we have access to huge amounts of information today, we are not very good at filtering or routing it in automated ways or between organizations. Web 4.0 will usher in an era of massive data flow and fantastic filtering tools to be able to aggregate, visualize and analyze data in real time.

As individuals, we want to be able to identify important information streams, combine, filter, and digest them into our minds quickly and easily. We are already starting to have a few online tools that help us do this. For example, IFTTT.com is a rules platform that lets a user pick a stream of information and trigger activities based on chosen conditions. IFTTT stands for "If This Then That." A basic rule could be if the weather is going to be rainy, send me a text message so I know to dress appropriately. We have a long way to go before we have consumer-level platforms for handling massive volumes of data, but the fact that we are building these tools at the enterprise level is a good sign they will migrate down to individuals.

So what will Web 4.0 connect? It will bring together data suppliers with data users so that knowledge can be expanded well past where we are today. We believe we have lots of knowledge today, but we know very little about why things happen around us. And we are truly blind to the vast majority of the Truth about our world. A manufacturer knows what it sold and seldom has any idea who bought something or what drove them to buy. A doctor makes a best guess at a diagnosis and rarely learns how a prescription really worked. A teacher educates kids and rarely knows the impact they had on those students later in life.

We seem to believe we know 90% of what there is to know, when we really probably know less than 5%. Web 4.0 will close that gap substantially and will provide what will feel like ambient intelligence, flowing around us all of the time.

Web 5.0 – Connected Humans and Technology (Transhumanism): This will not be the last era to be sure; it simply is where I will stop painting the picture for you. Web 5.0 will be defined by the tight integration of humans with technology at a physical level. We are on a path now to augment ourselves so that we can access what online tools are providing for us more easily. There is no reason to believe we will stop this march until we have interfaces embedded in our bodies so that we access the Internet, store data and process information within our bodies.

There is an argument that we are close to being an android already, if we define that as a man/machine combination. When I see people using a mobile device with a Bluetooth headset in their ear, it appears to me to be the first step. Add to this a smartwatch, and a heads-up display and we are now a couple more steps forward. There is no question that what we want is to have technology immediately available to us with the least effort necessary. We want it to be as simple as accessing our own five senses or brain. In order to do that, we will keep finding ways to embed technology onto - and into - our physical self.

As this happens people will end up being in various states of augmentation and this will have huge implications for the human race. In fact we will talk a lot more about this in later chapters. It is enough to say here that when we connect technology directly to our bodies, that connection will have a powerful influence on how we learn, how we solve problems and how we perform. It also will remove the last barrier to having a true Universal Mind, because anyone who chooses to be augmented will have instant access to the collective.

I will discuss later in the book the concept of Transhumanism, which is the idea that, at some point, we will be able to so augment ourselves with technology that we will be a different species than what we know as human today. The integration of technology into the body with Web 5.0 might very well be the tipping point where we can say we become "transhuman."

Most certainly there will be a Web 6.0, and we will see more powerful dynamics emerge from the connections that technology will enhance. Perhaps this will be a connection between people that is so tight we will have completed the creation of a fully functioning Universal Mind. Maybe we will gain the ability to connect through unseen channels that we believe exist yet have not really proven scientifically. Possibly our Web 6.0 level of connection will involve spiritual components that most people struggle to accept today.

Hopefully with the framework I just provided, you have a better picture as to how critical connections are happening in waves. Historians will be able to look at our present decades and show lots of interesting forks in the road for humanity. Since we are living in the middle of this quickly-changing world of connections, it is hard to appreciate how significant the changes through these five eras will be. Our job is to gain understanding of what is happening to us at each moment and be wise about what we connect to, and how.

Although machine-to-machine connections will change our lives quite a bit, I would like to focus the rest of the chapter on person-to-person connections: how we flow information and emotions to each other has everything to do with how we grow and prosper in our lives. Starve a prisoner of information and we know the results. Starve a child of information and emotion and we potentially wither not only that child, but their offspring in future generations as well. For this reason, we must do everything we can to ensure that our next generations have the healthiest and most nourishing connections possible in their lives.

When we the choose digital connection methods that we believe improve our lives, we do not realize often the collateral damage we are doing to our relationships. Let's take a deeper look at exactly what flows to each of us and how this provides value in our lives. I believe there are three categories of what flows to us when we connect electronically, and a blindness to this is what has dangerous ramifications for humanity at present:

The flow information: I define this as data, news and updates of status that are not of a personal nature. We can get a flow of information from teachers, television, books, Websites and other data providers. For our purposes vis-à-vis the human-to-human connection, the flow would be information provided by experts, business affiliates, parents, pastors, doctors, friends, and even other people or sources we're unaware of. We have dramatically improved this method with our digital tools because we have taken geography and costs out of the equation, creating an ability to "talk" to hundreds of people a day and consume information, provide information, or flow it bi-directionally in real time. This information feeds our heads much more than our hearts.

The flow of personal thoughts, opinions, and ideas: This flow is differentiated because, although it can be classified as more information, it comes from a place of personal experience or bias. This flow is permeated with more emotion either in how it is delivered or in the impact when received. This flow allows us to learn a lot about the person sharing this stream with us and, in turn, we can share much of ourselves by providing a flow of our thoughts, opinions and ideas with others. When we flow this stream through a connection with another we can touch both the head and the heart.

The flow of love (or hate): While the first two connections primarily are transferring words, pictures or videos to inform, this style of connection is dominated by flowing emotions, feelings and energy. While sending someone an "I love you" through your Apple Watch is a nice sentiment, it is not the same as saying it while looking your significant other in the eyes.

I can connect to someone by hugging them and holding them tight and this will pass love to them without saying a word. I can connect with someone by looking them in the eyes and smiling at them. I can also connect and flow the opposite to them by looking them in the eyes with abject hatred. For all of time we have had the ability to connect directly in a physical presence with others, and this form of communication gives us the ability to connect at all three levels. Face-to-face also is the only method of connection that allows us to connect at all three levels. Skype or any other video call system might allow us to connect face-to-face from a virtual level but we cannot pass any energy through a physical touch or physical closeness. Obviously, this level of connection is primarily done at the heart level.

Just to be clear, I am making no value judgment on the flows that reach the head vs. the heart. Both are necessary and both fill a need in us. We need information, ideas and opinions to help us become more enlightened. We need a flow of love to help us feel valued, comfortable and secure. When we are starved for information we stay ignorant. When we are starved of love we wither and die at a soul level. Regarding my earlier point about the impact on prisoners of being isolated, I am sure it is not the lack of information flow that warped them; it was the lack of human caring and connection at a heart level.

I believe human beings were built to feel love, flow love and grow in wisdom throughout our years. To do this we need all of these flows in our lives. We could "starve" without any one of them. Technology is improving our ability to open up the volume and frequency of our connections and, at the same time, is giving us the easy ability to regulate the flow from each of the three. If we make poor decisions on the balance of what we receive and send through these flows, we lower our quality of life becomes, and possibly that of the people around us. Because this has huge ramifications on us all, let's go one step deeper in defining the difference between connecting at the head and heart levels.

The Levels of Depth in Our Connections
There is a Humalogical balance that people choose when handling their relationships. For most people in developed nations who own smartphones, we connect directly on a regular basis with people in our immediate presence. When the relationship is with someone who lives more than a few miles away, we often

lean on technology to bridge the distance. In most cases, even with people close to us physically, our relationships are nurtured through an admixture of face-to-face, texting, email and social tools. When we talk exclusively face-to-face we are at an H5 on the Humalogy Scale. If we only communicate with someone through Facebook and never engage with the in person, we would put the connection at something like a T3 because the vast majority of delivery is through technology. So if I were to take a stab at the Humalogy range that most relationships stay in, it would be somewhere between an H2 and a T3. Most people do not relate to people – even close to them – exclusively by direct connections.

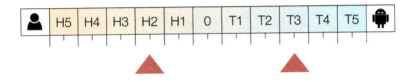

As mentioned earlier, we have a limited amount of time each day to invest in relationships. For that reason, it is logical to assume that, with the time we can invest, we have to choose between quantity or quality. This is just one of the hundreds of ways that technology gives and takes away at the same time. Through our online tools we can develop and nurture relationships with people all over the world, and for every minute we are doing that wonderful online nurturing, we have one less minute to devote to the people closest to us. For this reason, we need to be very conscious of how we invest our time in the relationships in our lives. Because all relationships are not equal in depth, we must factor in how many relationships we want to maintain at each level.

It is helpful to think about the connections we have between us as a series of layers. Each layer carries with it more meaning from a heart perspective. Each layer is important, and while they are very different as to how they feel, they all meet needs inside of each one of us. The critical distinction is to understand that we have a shrinking amount of people in each descending layer and for us to have a robust and meaningful life, we need to be very conscious of having a meaningful ratio in each level. To the extent that digital tools might be changing the ratio in the wrong direction, we must be aware of that trend.

Level one connection: Surface Acknowledgements (e.g. "How are you today?" "I am good, how are you?", "Good.") This is connection where we acknowledge someone specifically and really do not exchange information. Online, this would be like sending an email that says, "Just checking on you. I hope everything is good with you," or sending them a picture of what you are doing at the moment through Instagram.

Technology has expanded this type of relationship dramatically because social technology allows us to "friend," "follow," or "link" to a person who we might know casually, or not at all. By doing this, we can share whatever information we choose to submit publicly to them and thereby possibly deepen the relationship, or simply maintain it as an acquaintance forever. We know their name, what they look like and, to the extent they post anything, we know more about them than if we ran into them in person every once in a while. This is a head-level connection.

Level two connection: Information Exchange (e.g. "Did you see that the Yankees won 5-2 last night?" "Sounds like they played really well considering the weather.") At this level of connection, we exchange information with another person that may be valuable or not, and at least potentially we improve the knowledge level between two people. This connection could start with a question from one to another or simply involve sharing an opinion on a topic. In the digital world, this could come in the form of a Facebook or Twitter post, an email or through a video call.

This level of relationship is defined by having invested enough time with a person to have shared a good amount of information bi-directionally. Normally we describe this person as a friend, a coworker, a client, a fellow student, a person we have worked with regularly, or played with on a sports team. We have not shared with them too deeply about our personal life but have engaged with them enough that we have been able to learn more about how they operate and about their skill levels at whatever we are doing together at that time. This is a head-level connection.

Level three connection: Personal Experience Exchanges (e.g. "Last night I was walking my dog and I realized that I had not been outside for more than thirty minutes since winter began). At this level we exchange personal information about what we have done in our lives. This flows information that is specific to our lives so that another might know us better as a person. This is where we begin to tell stories and not just exchange facts. The digital vehicle for this could be email, Skype or, in rare cases, through Facebook if someone is an ardent chronicler of their life. This really cannot be done through Twitter, texting or pictures and videos because, although we can transmit the information on a personal experience, we cannot supply the color around it. We can send pictures of our vacation documenting everything we did, though that does not really transmit the richness of our experience even if we tag the photo.

The relationship at this level takes a bit of a turn because it will now "pierce the veil" as the connection ceases to be simply about information exchange or casual knowledge of a person. At this level there is the added component of sharing personal experiences outside of a family, business, or school setting. The veil of distrust is pulled away and we move toward a relationship in which we are willing to share observations that are very personal to us, as well as how we feel about them. People at this level share time with each other in an environment that is away from work, away from school, and is focused on some activity that is shared in their personal lives. It could be going to a basketball game together, going to each other's homes, or going on a vacation together. This is a head-level connection.

A Level Four Connection: Emotion Delivery (e.g. "Last night I found out my best friend from high school and her daughter died in a car accident. I have been sad and depressed ever since. She was such a beautiful person and I just don't understand why she would be taken from this world.") At this level of exchange, we are willing to let someone hear, see, and feel an emotion, and it is normally one party delivering their feelings while the other party might just be listening. The emotion could be joy, pain, fear, frustration, or anger. Normally we will be sharing information and emotion so we are making a more widespread and intense connection with another. The digital equivalent can only be done through a video call or possibly email if we are a good writer. I say this because video calls provide the visual and audio clues that we need as humans to really define emotions. Email restricts the emotional description to be written or surmised from the text. I suppose we could argue that sending someone a picture with a sad or angry look on our face or a really sharply worded text might get a point across, but that is not so much connecting through emotion as it is hitting someone in the face with our feelings.

The line that gets crossed at this level is that at least one party has to be willing to express their true emotions with the other. This means not putting on a happy face or trying to be professional. Instead we are more transparent and we express how we feel honestly to the other person. That could manifest itself as attraction or could simply mean we show true feelings that were considered too unbecoming to share at the earlier level. Sometimes we describe this as "getting real" with someone. At this level of relationship, we begin to transition from a head to a heart connection.

Level five connection: Flowing Emotions Bi-directionally (e.g. "I am so happy today because it is Friday and I have a great weekend planned!" "Same here; I am going to a concert and I have been planning this all year!") At this level the exchange is both ways, as both parties connect at a heart level and information level. So there is a flow of discussion and emotion going back and forth. There may be a problem shared, with empathy flowing back. There may be an emotion shared, with a caring response from our listener. This is a fairly deep and meaningful connection for us as human beings. We were designed to have and feel emotions; this separates us from machines of course. We can shove down, ignore, or fight emotions, but in the end we feel things and we are healthier when we feel and share how we feel. Once again this can be done partially through a video call, even though it is a weaker substitute for being present with someone. It really cannot be done through the many other connection methods we have developed, at least not in real time.

A relationship at this level is pretty rare for most human beings. This requires a high level of trust, empathy, and time. It is not possible to maintain very many relationships at this level because there are only so many hours in the day. The role technology plays in these relationships normally is one of staying in touch. People at this level care enough about each other to want to know on a frequent basis what the other person is doing and how the are faring. Technology facilitates this quite well because we can use it to check in many times a day, no matter where we, or they, are. Phone/video calls and text messages are a major piece of the communication system at this level. This is huge blessing compared with just a couple of decades ago when it was difficult to know what someone close to us was doing all day without being with them physically or reaching them on a telephone land line. This is a head-and heart-level connection.

Level six connection: Flowing emotions bi-directionally with a physical component (e.g. "I am sorry for your loss," and then holding the other person who experienced the loss.) This level is defined by the addition of touch between two people. Another reality for us as human beings is that touch matters. We exchange something powerful when we touch another physically. That could be love that is shared through a massage or hate shared through striking someone. In either case the message sent will have a powerful impact on the person being touched. Even being close to someone physically by sitting with them a foot away and looking them in the eye fulfills in us an ability to exchange energy at some level. This is where technology completely steps out because we cannot virtualize this kind of connection. It is the most powerful level when you think about the impact on our lives. It is also the level we are slowly depleting because of our technology use. If you doubt this, ask any grandparent who has sat in the same room with a grandchild who is addicted to their connections on their mobile device. This is a situation where a physical presence and exchange of meaningful connection is being decimated.

At the deepest level of relationships, we share our bodies with another person. We are willing to touch and be touched in intimate ways and to flow love to each other through this touch. All of the comments made in Level Five apply here. We do not have a technology substitute for the power of touch at this point. We can create a virtual world where avatars might touch people for us and we can even control devices remotely that will touch someone, but we cannot replicate the feeling and energy exchange of the human touch.

A level-six relationship is best represented today as a marriage or committed partnership. There is a deep physical and emotional connection with information exchange. This level of relationship also can be found between family members who are very close. This is a head and heart level connection.

Technology connects us at the head level while face-to-face interactions connect more at the heart level – we need to be conscious of the balance in our lives and to know when investing in each level is appropriate. For us to experience life in a fulfilling way we must find balance between levels one to three and four to six. We also need to develop the skills to be proficient at nurturing relationships at all levels. Too much time invested in one end of this spectrum or the other is not healthy and will lead either to many superficial relationships or an unhealthy attachment just to one. For many people, their first task simply is to be conscious of these levels and to seek a better balance between them.

Something to note through the various levels of relationships is how our choices of digital communication tools changes. When it comes to email there are levels at which it is natural for me to provide my address so we can communicate in long-form online. In order to text me you have to be at a level three and below. I will not video call with you unless you are a family member or really close friend. I will connect to most people on LinkedIn for business reasons but I will not connect with anyone other than family and close friends on Instagram or Snapchat. Over time we continue to mature our practices in this area as we gain new tools with more powerful capabilities. Constantly we are making decisions about relationships and whether they are "text worthy."

Another distinction is that, as the levels deepen, the Humalogy score moves more toward the human side. Said another way, when relationships are at a less personal level we will use more technology to facilitate the relationships. As the relationship deepens we will use technology to communicate with each other on a more constant basis but the technology plays a very minor role in building the ultimate close connection. In order to have deep and healthy relationships we need to be sharing ourselves with another human in proximity. There is no substitution at this time for that. Yes, we can employ video chat in a Level five or six relationship when we are apart but it is nearly impossible to build to that level without being in proximity for a decent amount of time.

Technology is having a huge impact on relationships in general, be they level one or level six. The basic dynamics that drive relationships can be changed because technology augments or alters the information flow between two people, and information flow is a large part of how we relate to each other. Here are two very different stories that show the difference technology can make in building or destroying relationships:

The first story: When I was a young teenager my family went to a church retreat of a sort in Ohio where I lived. I met a girl my age there and we spent a lot of time together that week. I really liked her in every way a 13-year-old boy could like a 13-year-old girl. She was pretty, she was fun to be around and she was smart. I remember being sad to have to leave her because she lived in a city that was a long way from mine and I knew I would probably not be able to see her again unless we all came back to the same place the next year (which did not happen). For months we wrote letters back and forth to each other and I was so excited every time one would arrive. She was the coolest girl I knew and I really wanted to be around her. The exchange of letters continued for about a year and then slowly they stopped. I am not sure who was supposed to send the one that ultimately stopped the connection but one of us failed to reach out and I never saw or heard from her again. I don't remember her name or even much about her other than the fact that I thought she was great and I loved getting letters from her so much.

Had this happened in this day and age with all our technology, the relationship could have taken a very different path. If I could have video called her every few days, emailed her, texted, or had her follow me on Instagram, we could have stayed close or even grown much closer. Technology could have augmented our relationship in a way that was not possible back in the 1970s. There is no way of knowing how things might have been different had we been able to connect as people do today, but my point has nothing to do with this failure to stay connected in my life. The point is that, every day in this world, technology is impacting millions of relationships and altering the course they take. Had I been able to stay in touch with her and grow our relationship I could have married her and my life would be very different than it is today: married to Annette. There would be different children in the world and the impacts ripple out from there.

I believe there is nothing that ever has changed the course of relationships quite like technology is altering them today. Who we meet, how we deepen relationships and how and why we break up all are being impacted at a rapid rate by digital tools. We have no way of knowing how the world would be different without the digital tools we use in relationships today, but it's fair to say they would be dramatically different.

The second story: Second Life is an online platform that allows people to create avatars and "live" in a digital world with just about everything that the real world has except physical touch. It gives us an early glimpse into what can happen with digital relationships when people have the ability to live vicariously through an avatar. Although Second Life has faded in popularity, I am very sure we will, once again, see a growth in people participating in virtual worlds in our future. In 2008 the following excerpt from an ABC News article hit the Web. Even though it was a number of years ago, this story shows an interesting insight into technology and relationships.

> *The line between actual reality and virtual reality has become more blurred with the advent of the popular online game "Second Life."*
>
> *The virtual world provides a place for individuals to create an avatar and engage in most everyday activities, including attending concerts, conducting meetings, meeting new friends, and apparently having virtual extra-marital affairs.*
>
> *Amy Taylor, 28, and David Pollard, 40, expect to have their divorce finalized next week. Their three-year marriage came to a crashing end after Pollard was caught by his wife e-snuggling with another Second Life female avatar lover.*
>
> *"I caught him cuddling a woman on a sofa in the game. It looked really affectionate," Taylor told Sky News.*
>
> *When Taylor confronted her husband about the matter and asked to see his chat history, he quickly turned off both the monitor and the computer to erase any evidence of his interaction.*
>
> *The incident was the last straw for Taylor, since it wasn't the first time Pollard had strayed digitally.*

Here is one of the first prime examples of technology facilitating a relationship that was at a fairly deep level. The only problem was that the relationship facilitated was in conflict with an existing marriage. This side effect is not limited to Second Life of course. Technology provides new ways for people to connect and this gives people many new tools to use in unhealthy ways. At the same time technology has the potential to allow a relationship to build where one simply could not have been developed in the past.

As virtual worlds add more capability to mimic reality technologically, they will provide incredibly lifelike experiences. If you doubt this, ask yourself why Facebook would invest billions of dollars in Oculus, a virtual reality company. Clearly they see that providing a very realistic experience that allows two or more people to relate as if they were in a different location was worth spending billions of dollars to acquire. As the virtual worlds we can play in become more lifelike, the potential to deepen relationships through them also will grow. This is a perfect example of technology that having dramatic capability either to move someone forward in life or drag them into dramatic conflict.

The purpose of this chapter is not to write the definitive anthology on technology and relationships, because that is a huge subject. What I hope to achieve simply is to get you thinking about some of the ways our lives are being impacted, and then as always, to discern whether these impacts are positive or negative for humanity. Now that we have talked about the types of connections and the various levels of depth, let's move on to a few completely new relationship/connection dynamics that have been birthed through our new technologies.

Online Relationships

The expansion of the Web created the whole possibility for online relationships. We have had long distance relationships for centuries and still do to this day. The only difference is that in decades past the long distance relationship also meant long periods with little to no communication. Today distance is irrelevant in how often we can communicate with a person because we have many new tools to stay in touch with anyone, just about anywhere. The distinction between an online relationship and long distance relationship is that the online relationship is not necessarily driven by the distance between two people. An online relationship speaks more to the preponderance of methods used to communicate. People can have an online relationship when they live in the same city if they spend the vast majority of their time talking to each other online.

There are some very positives of being able to maintain an online relationship with someone if there is no alternative or if the online relationship allows us to have a connection with someone that could not possibly exist any other way. A person who is disabled can connect with someone even though they might be bedridden or restricted to their home. A person who is very busy and can only connect with some people in their lives through the use of technology may benefit from the social tools online because they do not have the time to be face-to-face with everyone.

The very nature of an online relationship lends itself to allowing someone to build a bond with a person who lives far away, and this widens the potential pool of people with whom we can connect. Now it is possible to connect with someone across the country, get to know them and then take the relationship to the next level, or maybe even to marriage. There are many stories like this from the past ten years. At the same time, there are dangers.

There is a term for certain online relationships that has been popularized by the television show, "Catfish." Catfishing refers to the practice of hiding behind a made-up online persona while having a relationship with a person. This is on the extreme end of a spectrum. At the other end of the spectrum would be the little white lies people perpetrate with their online profiles, such as an altered photo that is just a bit nicer looking than we are in real life or stating that we have a college degree when actually we never finished. Both ends, and everything in between, are examples of a problem that can easily happen when using technology to augment a relationship.

When you are not physically in the presence of a person you are in the relationship with, it is possible to misrepresent the truth on something that would be obvious if you were in person. Being fair, even without technology people have been lying about their past or present in relationships as long as we have had relationships. Technology just allows us to provide misinformation at a greater level when people are not physically in the room together. Nowhere is this more apparent than in the online dating sites that have been growing as an accepted tool for people to meet each other. As of this writing over twenty million people in the U.S. have tried an online dating site and the cultural fears or embarrassment of this method are receding.

Although it might seem odd to people of the older generations, the concept behind dating or meet-up sites is really good and quite efficient. Instead of being forced to find a person to connect with (for an evening, or a lifetime) in person, by sorting through only those who are physically close, dating sites allow us to use criteria and profiles to find someone through the algorithms of the site. Over the past ten years we have seen fast growth and diversification in this field, as there is a growing number of dating sites and mobile applications that are specialized for our specific interests. There are sites that specialize in finding a partner for a one-night hook-up and there are Christian-focused sites to help find a like-minded spouse. There are sites that match up farmers, older folks, people who are near you physically at the moment and people in the LGBT world. These sites vary in how sophisticated the matching systems are, with one end of the spectrum being a simple self-selection by sorting through profiles, to eHarmony and Match who work hard to help us find the perfect mate through algorithmic matching. Various sources state that over 17% of marriages, or one in six, are currently generated through online sites.

Even at this rate we have to admit that the impact of dating sites in the U.S. has already had a massive impact on who is married to whom. Or looked at in the reverse, who would these people be married to instead of their current spouses without a dating site (if they would be married at all)? A substantial number of marriages are coming from these tools, and most of us have at least a few friends that met their spouses this way. Dating sites are a microcosm of the good and bad that technology often creates. On one hand, they help lonely or searching people by providing efficient and sophisticated ways to find someone to connect with. This can be a great thing because it widens the net so people really can find the right match. On the other hand, as mentioned earlier, they are a breeding ground for people to misrepresent who they are in order to trick another person into connecting with them. The motives behind the deception can be innocent or aggressive. These sites also can become a crutch for lonely and introverted people who would rather connect over the Web than develop in-person social skills.

I speak here as one who has never used a dating site, so all of my comments are based on observation and research. I have nothing against them; I just happened to get married before they become popular and have not wanted my wife to see me visiting them! I do want to offer one bright ray of hope on this subject. I recently heard the founder of eHarmony speak at a conference and he told the audience that the divorce rate of people married on eHarmony is only 3.8%. While I have not validated this if it is even close to that number over the long term, it is a good sign that, with the right matching schemas, we can use technology to help people choose better mates. I am all for anything that helps us lower the divorce rate, as I expect most people would be.

One last positive to mention: while I grew up in a large city my wife grew up in a very small town in Oklahoma. She taught me about the relationship dynamic of having very few boys her age to choose from in town. Therefore, people who grow up and stay in small towns have very few choices for a mate. This can lead to people "taking the best they can get" instead of finding someone who is a great long-term match. Dating sites provide an ability to expand our field of choices and there is nothing wrong with this. I have observed that many married couples who got married before the time of dating sites are curious if they would have chosen their current spouse had they found them on a dating site. This leads to wonder, "would I have found someone better online?" There is no benefit to ruminating on this too deeply because it can only lead us to invest less in what we have and more in thinking about what could have been.

Forensic Relationships

A close cousin to the ability to find new relationships through online sites is the ability of renewing relationships that would have been long dead. A forensic relationship is a term that some people use to describe this scenario. I have heard it estimated that each of us has, on average, twenty-five relationships that are in existence that we would not have had, were it not for finding someone from our past through searching on a social site. Most of us have had the experience of getting an email or being followed through Facebook by someone we have not seen in many years. In some cases, this is a really great connection that brings back a flood of great memories. In other cases, it is someone who we wish would not have found us.

24% of internet users have searched for information online about someone they dated in the past, up from 11% in 2005.

24% of internet users have flirted with someone online, up from 15% in 2005.

Source: www.pewinternet.org

As a communication flow begins to happen some of these relationships will rekindle. Memories are shared, statuses updated and, depending on the situation, a full-fledged ongoing relationship restarts. This can be a newfound joy that we never thought we might have. I have had people who I went to high school with connect with me, and just seeing their picture and exchanging a few words has meant a lot to me. My life is just a little bit fuller because I know they are alive, and that I can reach out and talk to them anytime I want. I have had people I worked with years ago reach out through an online connection like LinkedIn, and we have gone on to help each other in a business context in some way. In decades past these were relationships that simply would have disappeared from our lives because we did not have a way to easily find each other even if we wanted to.

Last year a woman reached out to me. She was someone I had not thought about for years and, at the same time, I could never forget. This is not because we dated or had a long and close relationship, but because I had treated her terribly as a kid and always have felt guilty about that. She lived one street over from me and came from a working-class family who were first-generation Italian-Americans. She was not the prettiest girl at her young and awkward age and would get teased for her looks. She was always very quiet and shy, and was as nice as a person can be under those circumstances. Yet on the bus going to school people harassed her horribly for no reason at all. When people say that kids can be cruel, the way people treated her is a perfect example. As an adolescent I felt some kinship with her because she lived near me, and she would try to sit near me at times or look to me to be her friend. While I did not tease her like the others, I also did not stand up for her. To this day when I remember her I am embarrassed that I did not do more to befriend her. When she reached out to me over Facebook, those memories immediately flooded back and they were not very good. I remembered the teasing on the bus. I remembered thinking she really did not deserve it. I especially remembered the sadness on her face as she was abused for no good reason. When she connected with me online I took the opportunity to share my feelings with her and apologized for not standing up for her when I could have. We were communicating online about something that happened 40 years ago and I thank God we got this chance. She was very kind in her response. We don't actively communicate at this point however I am thankful that I got a few moments of her time to connect so I could apologize for being so weak when I should have been strong. I can only hope that my words brightened her day in some way.

When we connect with someone from our past there also can be powerful ramifications that cause damage to current relationships. In a strikingly large number of circumstances, people seeking out old boyfriends or girlfriends and renewing a connection leads to an illicit relationship that destroys a current one. According to geeksugar.com, Facebook is cited in one in five divorces.

Not all of the divorces that Facebook plays a role in can be traced to forensic relationships with an old flame. Some of these divorces can be caused by ignoring a spouse by spending time on social sites or by communicating with a random person in an intimate way that is not acceptable to the other spouse. Many divorces are caused by people reconnecting with someone from their past. In some cases, there is an unrealistic memory of how great the old relationship was so both parties seek to go back to a fairy tale land of first love, only to find that it is not as they remembered. If they are married, just revisiting old feelings with another can be enough to destroy the marriage. In other cases people renew an old relationship and choose to end their current one, finding happiness in that way.

Restarting past relationships that we thought might have ended is a magical capability that technology can afford us. There are wonderful benefits to being able to reconnect, and stay connected, with someone we made an investment in years ago. It also is a potentially dangerous path that leads to the renewing of a relationship that would have been better left in the past.

Technologies' Unique Impacts on Relationships/Connections

Although technology gives us the ability to communicate in very efficient ways, it also restricts our ability to do high-fidelity communications with all the capabilities a human can use. Often we cannot see the other person when they are initiating the communication. We cannot always see or feel the meaning behind written communication. There simply can be a lot of context lost when communicating through digital tools.

This creates the potential for something to be taken in a negative light when they had no intention to be negative. When person answers a few texts in a row with one word responses while driving it can be is perceived as curtness by the receiver. Or someone may appear to be much more interesting than they really are because they work hard to craft online communications they know will appeal to another. However, in person they are a complete dolt. This begs the question: can a deep relationship really be nurtured through heavy use of online communication? Or will there always be a missing piece of the relationship that is a truth gauge of sorts that we can only trust if we communicate with a person directly?

I have relatives who live halfway across the world and people I do business with who live across the country. In both cases, for practical reasons I cannot see them very often, if at all, face-to-face. While I might communicate with them regularly, I don't have a sense that I know for sure how they are truly doing in life. The snapshots I get through our digital communications are pretty sparse for sure. This allows me to keep a relationship moving forward at a certain level but it does not allow me to know these folks at a deep level really. There is nothing wrong with this situation; it actually is positive that I can at least stay in touch with them. However, I am clear that technology only facilitates a maintenance-mode level of connection.

The reason I make this distinction is to help you think about how the Humalogical Balance also impacts relationships. In this case the more we move toward an H5, the deeper a relationship has a chance to go. The more time the relationship stays at the T4 level, the more there will be natural limits on the depth we can attain. When people choose to lean too far into the technology side of the Humalogy scale by choice and not by circumstance, they may have bad intentions or psychological issues. When people choose to lean more toward the human side they are following a more natural path for building depth of relationships (not that this is a guarantee of a healthy relationship).

As we have discussed earlier, technology can afford us more volume and efficiency in relationships, though it can rarely give us depth. This causes concern for the older generations, as they see our young people using technology at a high rate to relate to friends and family. Ironically their high use of technology can increase their connections with older relatives if they are willing to use the same tools.

When I am speaking about technology to large audiences, a common set of questions I get from people over fifty years old focuses on their fears about their kids and grandkids. They will comment on their seeming lack of interest in talking to people sitting right in front of them, or that they have hundreds of people they connect with online and very few that they interact with in person. They worry about the younger generation's ability to build meaningful relationships.

This concern often stems from a personal experience between grandparents and their grandchildren. The good and not so good aspects of technology are summed up nicely when viewed through the lens of an older person who wants to have a deep and meaningful relationship with their grandchildren. While the younger person has many new channels to connect with a grandparent, there can be a broken connection if the grandparent is uncomfortable connecting through these channels. Many older people have fond memories of spending quality time face-to-face with their grandparents. Maybe they even talked to them on the phone from time to time in between visits and sent some letters back and forth. It is hard for them to feel the same connection when talking to their grandkids through a social website.

Is this a problem with the technology just not being able to facilitate the same level of emotional connection? Is it that the content shared over the wire is simply more surface than what would be covered face-to-face? I believe it is a combination of factors, and my concern is that young people feel like they are doing something nice by texting their grandma every once in a while, or letting grandpa follow them on a social site. Kids sometimes believe they are building a relationship in this way because it is how they do it with friends at school. The problem is that much of the wisdom that elders would like to pass down cannot be shared easily through social media. Wisdom needs to be passed through stories with clear emotions attached and this is hard to deliver online.

With the impact of technologies on relationships we can easily see the advantages and disadvantage when looked at through the lens of the connection between our younger and our older generations. There is a danger that we all need to be aware of: the potential to lose some of the family tree wisdom that could be passed down. The Internet itself can be a wonderful storage device for knowledge. At the same time, we should not depend on it to deliver all of the worlds stored thoughts. There are some areas of wisdom that are more easily passed down from a grandparent to parent to child in person rather than through a Website.

There is one final thought I want to cover when it comes to a unique aspect of technology and connections: the dynamic of connecting "one-to-many." There are huge benefits in the ability to connect at levels one, two and three on our depth scale with millions of people all at once with a keystroke. This means we have transitioned from the starting place of a person talking to crowds verbally on a hill, or in a square, to writing and printing an opinion and then fighting for distribution. Then onto using radio or television to deliver a message to the many, this being controlled by whoever owned the media. The Internet freed us from the costs, or choke points, of the media, and now allows anyone to "broadcast" their ideas or opinions across the world. We all decide what we think is valuable to read, so we choose not only based on what we will read, but also by what we pass around to our friends. This is a process of the crowd voting for what we think is worthy of our time to digest. The glory in this new one-to-many connection method is that it is open to almost anyone. We all choose who we will listen to, which means we might listen to a blogger instead of the mainstream media. That is a very democratic way for connections to happen.

This is an amazing capability we already are taking for granted because we quickly forget that, just recently, we did not have the ability to "talk" to the world at will. Anyone with an Internet connection has an online voice and may talk to everyone else who is connected; we are seeing very important impacts of this capability every day.

Obviously having the Web as an online megaphone amplifies both the good and evil that someone wants to put out in the world. Thankfully to this point it seems that we have more positive than negative messages being broadcast, along with a growing intolerance for the kind of negativity that comes along with online "trolls." I choose to believe that this is a reflection of more people aligning with the light and propagating the positive they find online.

One proof point of this is the trend toward what makes a viral video have a high pass-around rate. If we were to look at the last 100 videos that went viral on the Web we would find that the majority delivered positive messages. We actually do not see a high rate of negative or evil videos getting passed around in the general public. Certainly we have child pornographers who will pass devastatingly abusive videos of children to each other, but the general public does not align with this. We see "fail" videos that show people being injured through accidents and it is sad that people enjoy the misfortunes of others. Thankfully that is about the most negative type of video we see get high pass-around rates. I think this is a great sign for humanity at the moment.

Because of our newfound ability to connect one-to-many, the Internet has provided us with an ability to aggregate the minds of thousands of people at any one moment on a task, topic or problem. Sharing information with the Internet crowd is one thing; requesting help from the crowd is a completely different dynamic. We have already begun to swim in that ocean with crowdsourcing and all the other crowd dynamics now manifesting themselves. One-to-many is another extremely powerful type of connection that certainly will certainly serve us at the head level and, in some cases, allow us to touch the hearts of the collective world.

As we gain the ability to share emotional and uplifting stories across the Internet we will naturally increase the flow of positivity. I have noticed more articles and videos online that show wonderful acts of kindness. I have seen that the pass-around rates for this kind of content grow. I am hopeful that the vast majority of us will choose to connect with content, news and posts that are uplifting, and that this is what we want to amplify by sharing with others. With a medium that allows us to share one-to-many, if the majority choose to connect with things that are fulfilling and healthy the positive impact could be rapid and expansive. At a practical level this would mean choosing to flow things that are loving and not destructive. If love can flow this way throughout the Universal mind, we have to ask how close the Internet is coming to being a manifestation of God.

The logic behind this thought would be: If God is love, and love is delivered through connections with each other, and the Internet is the most powerful facilitator of mass connections, and those mass connections can flow love to any one member, even anonymously, then maybe God guided us to invent the Internet to help with the flow of love and goodness.

The Future of Technology Augmented Relationships

As we develop new devices and more sophisticated software we will continue to change how relationships are discovered, developed and nurtured. Augmenting relationships is big business because we as users are willing to pay lots of money and invest huge amounts of time to improve our relationships. For this reason, there will be many different organizations seeking the holy grail to provide us methods for being virtually present to participate in a relationship, even when we are thousands of miles away. We crave any tool that will help us have deeper and more meaningful relationships when talking to someone in person is not possible.

Holography and virtual reality will improve our ability to seem like we are present even when we are nowhere near. This ability to simulate our presence anywhere in the world will improve our sense of what the other person really is saying and means. Not only will we improve the ability to see and hear the full context of a person being with us; it also will allow us to record our message and send it later if we cannot be with someone in real time. This will allow us to provide a more robust message because body language and non-verbal cues provide a much richer message than an email or voicemail.

Websites that provide relationship matching services will continue to improve their ability to assure that people are being honest with their profiles and will improve their algorithms for matching mates. They will learn how to enforce consequences on people for misleading or mistreating others. As these sites continue to improve more people will trust them to find connections and, ultimately, spouses. In the future it will seem very normal to start a romantic relationship online. I am sure that the majority of people who start a relationship will be using some kind of online component to find and vet their dates.

Virtual worlds will creep back into day-to-day life. They will be very different from Second Life in that we will start using them both at work and at home. Gamers today already spend millions of hours inside virtual worlds while being represented by avatars. Their simulated worlds will continue to get more realistic and the tools they use to "see" these worlds will as well. Heads-up virtual reality (VR) visors will improve to the point that people will suspend reality while in alternate worlds because what they are seeing will fool their brains into believing they are in a different place. At the same time, these visors will migrate their way to offices where they will be used to allow an employee to "attend" a meeting in a different city, and experience the room and people almost as if they were present.

Finally, VR visors will be used to communicate between people building relationships to better simulate a direct visit. Social sites will integrate VR technology into their offerings so that, with the click of a button, we will be able to talk to a person or a group of people as if they were in the room with us (Facebook + Oculus). Because the level of quality, the VR devices will provide such a realistic experience while also providing the safety and convenience of not actually being with someone physically. Because of this most people will choose to start relationships in this way. Going on a live date with a stranger, much less a blind date, will seem quaint and outdated. We will no sooner meet a stranger in person for a date than we would send our kids out into the neighborhood without adult supervision. Users will easily make conscious decisions as to when communicating on a subject warrants a human visualized delivery, or a text-based delivery.

And with all the advantages of an online VR date we will learn that there is value in touching, tasting, and smelling someone that simply cannot be duplicated with all of our technology. Even when realistic visual capabilities make us suspend the reality that a person is not standing next to us, we will still know there is a difference. This will not stop some people from becoming addicted to relationships in virtual worlds because, in many ways, they will be less "messy" than our real world. In a virtual world a person does not have to deal with projected emotions at a deep level because the energy behind an emotion will be diffused and we will not feel the energy coming directly from the person. A person has complete control over walking away from any interaction online. Connecting with people becomes simpler than in the real world because being rejected by a new contact will not feel quite as harsh as an in-person snub. And best of all, people in a virtual world will be able to create a persona that will be somewhat different than they can portray in person. These aspects will make VR relationships seductive even if they are less real.

Introverts especially will be enticed to virtual worlds because it will not be quite as exhausting as dealing with a group of people in real life. The quality of a relationship built on proximity and one built online will blur further, and that will be great news for some people. For a handicapped person this will be a Godsend. For someone with a lot to hide this will become another vehicle for deception.

Addictions to VR aside, I believe that most people will become conscious of finding the proper Humalogical balance in relationships and it will become very normal for folks to know when they need to be in proximity with a person vs. connecting through digital tools. This will be true for grandparents, spouses, business people and other relationships. We will cease to be so intrigued with the newness of our digital tools and learn to use them at the right times, for the right things. Even as they improve with in quality and multisensory capabilities, we will use them a bit more discretely. I say this because the human race does a great job over time of finding how to behave in the most effective ways. There always will be people who abuse technology tools in dark ways just like there are people who abuse drugs and alcohol. The majority of the world, however, will learn to use technology to help build relationships in a healthy way.

Our newest digital connection tools like mobile devices, video calling, text messaging, and social tools have provided an inexpensive and convenient way for family members to keep up with each other much more often and in a more immersive fashion. When compared to mailing letters and talking on the phone on holidays, which is how I connected with my grandparents, we have much better tools. In some cases, this has helped bring grandparents closer to their kids and grandkids through this is fantastic new capability. At the same time, we do have some grandparents who just do not feel comfortable with technology, having not grown up with it and because of the complexity of learning it. So they just opt out of these connection paths altogether. In both of these situations the older generation has made a choice and that is fair enough. There is one other scenario that scares me and that is where the older generation can, and does, connect through technology but the relationship stagnates at level one or two because neither party can relate to the other.

This can happen easily when grandparents "follow" their grandchildren on social sites, post comments online, and send texts and emails once or twice a week. The volume of connection between them may be greater than years past. However, the depth of the conversation is not. The kind of historical information and wisdom that grandparents shared previously had to be handed down in person and through an oral history in order for it to be rich and meaningful. It is nearly impossible to get this kind of connection through texting and Instagram posts. Here is a story from my life that illustrates this:

When I was young, I would spend the day with grandpa Nabb making sales calls while he drove all over northern Michigan. He sold printing and envelopes to banks, hospitals, schools and the like. He would go to a number of the small cities, call on his handful of clients, check their inventory, take orders, and move on to the next town. He drove an Olds '98, wore a fedora and smoked a cigar. He drank martinis dry and was the prototypical salesman of the day. When I visited their house I would go with him in my little suit and make the calls. This was an exciting adventure for me as a young kid and I really looked forward to this because I got to spend one-on-one time with my grandfather. We would eat out at restaurants that only a peddler would love, talk for hours while driving between towns and meet lots of businesspeople who had become friends of his. He bought me donuts in the morning and normally some kind of gift my parents would not buy for me. On a few trips we had to spend the night at a roadside motel because we were too far away from home to get back. He chewed wintergreen Dentyne gum to hide the cigar smell. I could tell you every detail of those days we spent together. He taught me how to sit properly in the chair across from the client. He taught me how to be charming when talking to them. He told me endless stories, Polish jokes (because my name was Polish in his mind) and taught me about being a salesman while we were on the road. We connected at a deep level during these trips because of the hours we spent together without distraction. I mourn for some of our young people today because they would spend most of their time on trips like this connected to their friends over mobile devices and paying little attention to their grandfather. They would hardly be able to sit still in a business meeting without engaging their screen because being bored even for 60 seconds is annoying to them. The world my grandfather worked in is pretty much gone. We don't sell printing that way anymore and normally we don't wear three-piece suits and fedoras on sales calls. Knowing Grandpa Nabb, had I owned a smartphone in those days he would have either taken it away, left it at home, or thrown it out the window.

The disconnection that is evolving between the generations in some families sets up a situation where older people cannot earn respect from the younger for the wisdom they have gained. If society loses much of its ability to pass wisdom from one generation to the next, we will have set up a situation where the young have to learn every lesson on their own. Take a look at what is happening in your own family regarding time spent face-to-face and sharing stories passed from your oldest generation to the young. My stepfather had long-term Alzheimer's and ended his life in a fog that was virtually impossible to penetrate. I saw firsthand how before we knew we had lost something, his memories were gone, never to be retrieved. Multiply this kind of loss through disconnection by millions of people and it becomes obvious that if we lose an ability to transfer wisdom from the older to the younger we all will suffer.

A trend we need to be thoughtful about is the dichotomy of having an awesome ability to connect instantly and in so many different ways at little cost anywhere in the world vs. feeling less connected. Being a technologist by trade I am the first to try every new connection method. Being an introvert I am a huge lover of using technology to talk to people so I don't have to feel the impact of being with lots of people face to face. Having an efficiency gene, as my wife labels it, I like to multitask connections so I can get more in during the day. I wake up every day fascinated by, and addicted to, digital communications.

As an observer of what technology is doing to us I have grave concerns about the disconnection of any generation because of technology proliferation. Today the instigating tools might be mobile devices, texting and social technologies. Tomorrow it might be brain-computer interfaces. Regardless of the cause, if an older generation is blocked from connecting with the youngest we will be diminished collectively for it.

What we do know for sure is that the Internet has exploded the number of connection tools available to us and we need to be aware of, and wise about, how we use these connections in our lives. We need to be aware that we can connect at the level of head and heart, and that either one does not replace the other. The Internet is much better at connecting us at the head level and being with someone through a direct discussion makes it easier to connect at a heart level. As we are becoming enamored with our digital tools we must not lose the value we derive from a real hug, a laugh out loud shared in person, and stories that get told for hours from one person to the next, and even from one generation to the next.

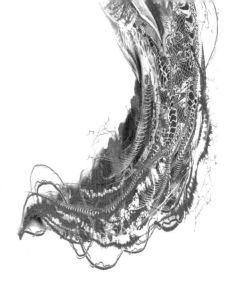

SUMMARY:

Digital Connections, Relationships, and the Changing of Tapestry of Life

Technology enhances our ability to connect and build relationships. This is nothing new to anyone living in our world today. What we do with this new power to connect is what weaves a new tapestry in our lives. Online connections give us the ability to add many more people to the fabric we are creating. The connection with devices makes it easier and more convenient to get things done. It is difficult at this point to imagine what our lives would be like without the impact of our current - let alone our future - abilities to connect with the people and things around us. Frictionless connections through technology, however, also can lead to trading the volume of connections we have each day for the depth of a personal connection. In that case we will nurture the mind but not our hearts. We must be thoughtful about the role connections play in our lives and how we choose to invest our time in them.

We must also be very thoughtful about using technology to enhance relationships where possible, while also recognizing that connecting someone directly has additional layers of value. We must understand that everyone around us might be in a different place when it comes to using technology to communicate and we must not force our ways of communicating into a relationship. That means not expecting someone to text message with us if that is not their strong suit, or limiting one's contact to face-to-face if the other party relies on technology as a primary connection point at times. We must adjust our preferences to fit the specific relationship.

We must recognize that we have new avenues for relationships through the ability to nurture connections with a large number of people all at once. A blogger who speaks to their community can connect with people virtually through their work wherein the entire basis of the relationship is the love of a specific hobby or interest.

And we can find a mate by using a dating site or application to help us match up with someone who might be available to get to know us.

With all of our newfound power to connect we must be thoughtful about what we are putting out in the world. If we create and pass along content, comments and posts that are uplifting, helpful, and constructive, we will be using the overall Web of connections to flow love to the collective. It may replicate out in ways we will never know.

My light-vs.-dark ratio for technology-augmented connections and relationships having an impact on humanity is 65% positive and 35% destructive. With that said, the Internet can amplify any state of being in an instant. As long as the majority of us chooses thoughtfully what we are providing when we connect with each other digitally, we can amplify the good.

Humalogy Viewpoint

As our ability to connect and build relationships continues to evolve we will need to find the appropriate balance between using technology and connecting directly with people. There is no easy way to choose a general score for humanity because we have people who are at both ends of the spectrum and wish to stay there. My best generalization is that younger people tend to be at a T2 with how they connect with people and the older generations tend to be at an H2 or H3. It is very possible that the best place to be will end up at a 0. This means that there is an equal balance of connecting with people through technology and in person. This could give us the best blend of efficiency, scalability, and depth.

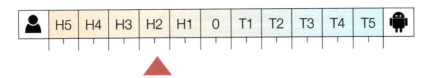

There is no question that there will be a significant amount of personal preference for where a person believes they should fall on the scale. Perhaps more compelling though is the question of what will be best for us at an emotional and spiritual level. Too much technology and we lose the depth of connection; too little and we lose the ability to reach across time and space to nurture relationships with people who are not close to us.

Seminal Questions

Are we trading how we invest our time by nurturing digitally a larger number of level-one and level-two connections instead cultivating deeper and more meaningful connections, thereby trading quantity for quality?

Instead of connecting us, is technology creating a chasm in our families between the younger and older generations?

How has the ability to connect with the larger world changed our lives? Are we using this new capability to share our gifts with the world?

With all the technology we are using, ultimately do we feel closer to the people around us or further apart?

When every device around us is connected, intelligent, connecting to us and passing data to other devices, will we have created an electronic partner for humanity or a crutch that will weaken us?

Ultimately will we use the ability of the Internet as a connection system to flow love and goodness around the world and uplift the planet, or will we miss this chance to become more enlightened as a species?

CHAPTER FOUR:
Information Immersion

__Data__ is a powerful raw material,
__Information__ can enlighten us,
__Knowledge__ helps us prosper,
while __Wisdom__ can save us all…

In the last chapter I mentioned that information is the first critical element that flows within the connection of people or even among machines for that matter. Obviously gaining information and knowledge has an intense impact on our lives because the more we have, the better decisions we can make. The better decisions we make, the healthier we are and the more we prosper. Because I am a technologist I would love to approach this from the angle of Business Intelligence and Big Data for the next twenty pages. However, this is not a business manual and, although I love the power of data as a business topic, I want this chapter to be meaningful to non-business people as well. For that reason, I am going to use plain language and forego talking about operational data stores, data warehouses, MDM, ODBC, SQL databases, ETL and API layers (note: I did not even bother describing these in the glossary so don't go there). These are the kinds of terms data specialists use everyday to describe the work they do.

Let's take a look at the power of raw data and the ways we convert it into various things that are more valuable from a technology and spiritual viewpoint. As strange as that might sound to some people there is an obvious connection between our unique (among animals) ability to synthesize data and our need to seek a higher calling. We use data and information as the bases for learning and if we continue to learn and grow as a species surely we will become more enlightened as we mature. This is a powerful idea because making wise choices and becoming more enlightened allows us all to live in a more peaceful and joyful state. If this sounds overly utopian to you then consider the alternative. If we were to revert to being crude and uneducated versions of ourselves, we would also have very short and painful lifespans without much love or joy flowing between each other. There are two directions we can go as humanity: forward toward enlightenment or backward toward self-destruction.

This will be dictated principally by how we progress with the process of learning the Truth about the world. The more we have the ability collectively to learn the Truth about why we are here, how we can bring our gifts to this world and how we all can live joyful and healthy lives together, the better we all will be as a species. The DIKW Chain is the progression of Data, to Information, to Knowledge, and finally to Wisdom. It simply is a model to use so that we understand the process of learning, problem solving, decisions making and creativity. Technology is providing wonderful tools to help us improve on all of these, and that has massive implications for our future. Data and God are not two completely different entities with nothing connecting them. God gave us brains that can utilize data and information to grow and mature. He did not give this to any other species. What we do with data and our new tools is up to us; it is our free will.

Let's start with a practical discussion of the DIKW Chain in case this is a concept you are not familiar with. This diagram will give you a quick visual to understand the concept:

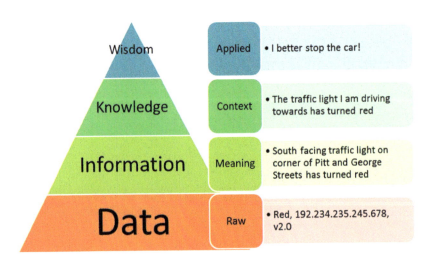

These can be looked at as ever-increasing levels of value that stand on the shoulders of the layer before. So data is a powerful raw material and is pretty useless to most people until it is organized into digestible information. Information is helpful, but until it is moved into your mind, put in context, and verified, it is not useful knowledge. You can have all the knowledge in the world yet, unless you apply it to make wise decisions, you are little more than a computer holding vast amounts of knowledge without the discretion to know how to use it. In my business world we strive to help companies do everything they can to master the DIKW Chain because using data well in the market results in better visibility into the Truth and the Truth can make you money. It also can set you free on a personal level.

We live in a world that is being impacted in historic ways because we have an explosion of data and information we can access now from anywhere, at anytime. Being able to tap into the Internet and all that it holds is something that science fiction writers wrote about for decades, yet many people never believed would happen. We need to see clearly what this is doing to us and where we are headed as we continue to improve this capability. Each of us, hundreds of times a day, sort through data that either is pushed to us or that we harvested, then we move it to information and finally imbed it in our brains as knowledge. This model explains that, as we get older, we get wiser because we have had more chances to gather data (by reading, listening or through experiences) and move it up to knowledge and wisdom. We hope to retain it, apply it and ultimately make great decisions.

Here is a critical distinction to understand about all this...

> *For better or worse, the consumption of media, as varied as e-mail and TV, has exploded. In 2008, people consumed three times as much information each day as they did in 1960. And they are constantly shifting their attention. Computer users at work change windows or check e-mail or other programs nearly 37 times an hour, new research shows.*
>
> *The nonstop interactivity is one of the most significant shifts ever in the human environment.*

Source: Adam Gazzaley, a neuroscientist at the University of California

The first two steps of the DIKW chain (generating data and organizing it into information) can be done by technology without our help. Humans generally own the second two levels when it gets to knowledge and wisdom. I will stipulate, however, that software also can approximate our ability to apply knowledge through the development of code-based rules. By mimicking the processes we would go through in our minds of using the information we are holding to make decisions, a computer also can come to the same conclusions based on variables presented. We call this kind of software Decisions Support Systems (DSS). I mention this because machines are getting smarter through copying how we think and process in order to make knowledgeable decisions.

Even though machines are growing quickly in their ability to manage data, information and even stretch into DSSs, it does not mean that humanity will make wiser decisions because of our computers. We might make better and faster business decisions by augmenting our own minds with computers but, at a personal level, we might operate differently because what the data shows might be in conflict with how we feel. We certainly have more help now in gathering information into our brains than at any time in history. We could even argue that much of that information actually gets digested as knowledge. None of that, however, means we will make smarter decisions.

Maybe an example would help to show the distinction I want to make. We have an amazing ability to stand in the middle of a city and have our mobile devices find any type of restaurant we desire. We also now read through the ratings and recommendations from lots of people who have eaten there. We can look over their menu online and even make a reservation, all while sitting in a car at a red light. This is an amazing amount of information at our fingertips that we did not have ten years ago. If we go to the restaurant and eat a 3,000-calorie meal filled with fat, salt, sugar and chemicals, then wash it down with another 1,000 calories of alcohol, we have just proven that we can have huge amounts of information on a subject and still make unwise decisions.

For our purposes in the rest of this chapter I am going to generalize with the term "information" as the way to signify the combination of data and information.

Years ago, information was stored in libraries, which primarily were large buildings with shelves full of books. To find information we searched for it through looking in a big wooden box with little drawers full of index cards. It was a slow and laborious process to harvest information and the quantity of information was limited to the volume of books provided in that building. Librarians made everyone be quiet because it took significant amounts of concentration to find the information we were looking for while poring over chapters. Frankly, I do not miss this experience.

Do we really appreciate how powerful it is that now we can search for any piece of information *in the world* and have it at our fingertips in seconds? Step back and ponder how different this is for us today. If you, like me, lived in a time when information was a lot harder to come by, you might have the same sense of wonder at the instant answering of just about every question. I believe we are going numb to the fact that this is a world-changing product of technology already. The simple ability to search the Internet and "pull" information to us is becoming so routine that we don't even notice or understand what goes on behind the curtain. It has helped create the magic box in our hands that answers billions of questions every day.

We ask a question of a search engine and, instantly, multiple screens full of possible answers come into view. Very few people in the world have any sense of all the technology involved in that simple search function. More interestingly, most people could care less how it happens, they just like the ability to ask a piece of technology a question and get an instant answer. From all over the world, the search engine collates sources to help solve problems and add to our knowledge instantly. In fact this has been a huge step toward being able to tap the Universal Mind of collected people. So again: do we appreciate the wonder we have created?

The opposite of the pull capability of searching online is a "push" capability where information we are interested in is sent to us in a constant stream. We simply need to choose the application we want to use to digest this river of information and we are instantly tapped into more than we could possibly consume. In my world I carry multiple devices; each one harvests a different type of information from the Web. I digest this information at any moment when I am not actively engaged with something else. The reason for this is that I love to learn, so immersing myself in blogs, newsletters, tweets, PowerPoints, infographics, news streams, white papers, videos and posts from friends and thought leaders has changed my life. Twenty years ago I could absorb only as much information as I could get from books, newspapers, TV, magazines, and other people, and most of that was not in real-time. Today I can tap into many more sources, 24/7 in real time and at a volume I could not have gotten in the old days even if it was all I did all day.

I spend my life giving speeches all over the world on the topic of digital transformation. My topic changes about every thirty minutes as you can imagine. When someone is a former pro athlete, a politician or a disabled person who has overachieved what normally could be expected, they have a shot at becoming an inspirational speaker. In their case they can give the same speech over and over for twenty years. Because I am a content speaker and my subject changes so fast, I am forced to give a new speech just about every week. At a minimum I have to update my last presentation constantly to be current. In order to pull this off I have had to build a powerful river of information. The concept behind this is that I use online tools to harvest information on technology and aggregate it to a few screens. Then I digest it every day for at least thirty minutes. In other words, I pour vast amounts of information into my brain every day. I absorb vast amounts of information compared to what I could digest ten years ago. The Internet now provides an unimaginable volume of information on every topic, and it grows geometrically. I carry a laptop, tablet, and a mobile device, and I absorb information from all three every day. Early in my career when I used to do Q&A after a speech, sometimes people would ask me questions about technology that I had no answer for, often because I had no knowledge of the subject they were asking about. Today I rarely have that problem because there is nothing someone asks that I have not been exposed to through my river of information. I stand in a much bigger river of information at this point and, thankfully, absorb a good percentage of it.

Before I dive into discussing how our new relationship with data and information is changing us, I must share some obligatory statistics at least on how fast the volume of information to which we have access is growing. These are just a handful of metrics I have seen lately, and they give us a sense of how our creation and archiving of information is exploding:

The amount of information in the digital universe would fill a stack of iPad Air tablets reaching 2/3 of the way to the moon (157,674 miles/253,704 kilometers). By 2020, there will be 6.6 stacks.

Source: "Executive Summary: Data Growth, Business Opportunities, and the IT Imperatives." *The Digital Universe of Opportunities: Rich Data and the Increasing Value of the Internet of Things*. April 2014. EMC Digital Universe with Research and Analysis by IDC.

In 2017, the gigabyte equivalent of all movies ever made will cross the global Internet every three minutes.

Source: "The Zettabyte Era - Trends and Analysis."
Cisco VNI: Forecast and Methodology, 2012- 2017.

So obviously we are in a time when the volume of information we create, consume, and store is growing dimensionally. To put this in a bit of historical perspective, let's look at a few large-scale growth spurts in our ability to create and store data using computer technology over the past fifty years:

1. Mainframes were the first serious computing devices that gave us the ability to store what, at the time, seemed like a large amount of data. Not only did we invent the ability to do machine based computation; we also created new storage media to hold the software code and data involved in the process. These magnetic-based storage systems were based on large tape reels and, later, disks, and platters. Although we would think them crude today, they represented the revolutionary ability to store large amounts of information on something other than paper. This era also introduced us to the ability to search and find information quickly from a large base of data.

2. Personal computers came along in the early 1980s. Within ten years they expanded the volume of information in the world dramatically in two ways. First, they gave the ability individuals to store data on their own hard drives. This meant that the millions of people at that time who owned one began filling up whatever drive size they could purchase. Then, like now, as soon as manufacturers gave us a larger amount of storage, we found data and files that needed storing. Second, PCs gave individuals a new tool to create the information; and create it we did. We built spreadsheets and wrote documents. We stored contact information for our network of friends, family and business associates. With our newfound ability to run whatever software we chose for our PCs, we turbocharged the volume of what was created, stored and shared. The sharing of information was clunky because we had to

copy it out to a physical disk of some sort in order to hand it over to someone else. That would change soon.

3. The Internet/Web provided the transport protocol to be able to share the information on our PCs much more easily. Once we were connected to the Internet, information could be shared through direct file delivery or it could be made available for access through Websites. Search engines made it possible to go find information from anyone who had posted something, anywhere in the world. This resulted in a high motivation to create information and provide it to others in order to promote our products, services, ideas and opinions. Soon after using the Internet it became obvious to us all that having access to a vast collection of information was a valuable thing, so billions of people now add to our collective database.

4. Cloud computing and social technologies give us an even easier ability to upload mass volumes of content in order to share it with each other. Remember that there would be no Facebook, YouTube, Instagram, LinkedIn, Twitter, etc. without the cloud. All of these sites provide their service over the Web without the need for us to have a server to run them and all their information storage is done in huge data centers that are not in our houses. The volumes of information that sites like this now handle is staggering. For example, more video is uploaded to YouTube in 60 days than the three major U.S. networks created in 60 years previously (From Google's statistics page).

5. Mobile devices and Internet of Things (IoT) have exploded the volume of information yet again because information such as location (through GPS) can be harvested and stored among billions of people, moment-to-moment. The IoT capability of sensors being built into many physical things is growing rapidly, so information on everything from temperature changes, car speeds and our health statistics are being captured and stored.

As we have been taking these steps forward, technology firms have been lowering the cost to store all of this information and accelerating the transfer speeds for gathering and delivering it. Best of all, we are not finished making improvements to how we handle data; in a sense this is only in the embryonic stages. We will continue to harvest and gather more information, and speed up the processing and delivery of it to whomever (or whatever) needs it. We now harvest data at a scale that is mind-boggling. But we have a much greater ability to create and store data than we do to turn it into useful information and knowledge. It is likely we will spend decades learning to extract the value and insights that hide in our computers today.

This has thrust us into a new world of possibilities where we need to be thoughtful about how each of us is handling our own personal DIKW chains. For each of us there is a lot at stake. Knowledge is a powerful force in helping us become enlightened, prosperous, and healthy. In fact, the DIKW chain is the only way to make improvements in these areas. If we are to make great strides forward collectively, here are a few areas that will need to be addressed:

Throttling down the flow of information so it is manageable: Building the ability to tap into an overwhelming river of information can be both a blessing and a curse. Because we have applications that allow us now to connect with streams of information from many sources (e.g., our work, friends and family, news reports, social sites, subjects of interest) we can set up a situation where it is impossible to keep up with the flow we have assembled. A good example of this is our social stream. I am connected to thousands of people on Facebook, LinkedIn and Twitter, and it is impossible for me to fit in the time each day just to monitor the stream of information all of these people post. I have managed my river of information by throttling it down so that I can focus on my priorities first, then the "nice to have" information later. I did this through a combination of how I organized the applications I use to access the information and the habits I have formed for investing time in this practice. If I were not to throttle down these information streams appropriately, there would be consequences to my performance in life. We all will have to grow our abilities to restrict the flow of information so it does not overwhelm us. This leads to the next skill.

Filtering the flow of information so we can digest the important aspects: In addition to throttling the flow of information coming at us, we need the ability to filter our rivers of information so that the important bits float to the top for easy access. In truth, the software/application industry is not very advanced with this functionality yet. We are still at the stage where companies are excited to create huge flows of information and provide it to us, regardless of the value. Keep in mind that soon, our rivers of information not only will be coming from other people; because of the Internet of Things, we will unleash all of the devices we own to talk to us as well. There also is a good reason why companies have not focused on giving us filtering capabilities just yet. Honestly they don't really care that much about helping. Many of the tools and apps we use to access streams of information either are sold on a subscription basis or are advertiser-supported (e.g., social networking sites). The subscription sites probably believe that more information is a better value for the monthly fee, so they pour it on. The free social sites want to make money from advertisers so their motivation is to get more eyeballs using their service and not to make the information they provide easier to filter. In fact there is a negative incentive because if we spend less time looking at our "friends'" posts, that hurts the companies' advertising metrics. We must improve the ability to give high value weighting rules to every information supplier so that the pieces we really want the most all float to the top of our information feeds. Essentially, we need much better tools that allow us to personalize the relative importance of all the information we could have at our disposal. Then we will each need to build our own filtering systems so that we get more of what we really need and less of what really does not make a difference in our lives.

Balancing the digestion of information with contemplative thinking: One of the insidious things that I believe is happening to some of us is having access to more information while converting less of it to wisdom. In the old days people had a good excuse for being ignorant; they might have had very little access to the information that could have enlightened them. Today we have no excuse. This dynamic is happening for a few different reasons. People have access to unprecedented amounts of information at their fingertips, and they know it. So if they need information they can grab it, use it and then forget it. They know it always will be available to them again in the future. Subconsciously they make a decision not to move information to local knowledge in their own brains. Add to that the dangerous potential for people to start relying on the Internet and its information to solve every problem for them. This could lower the "muscle" in our minds that fights our way through situations to a solution or memorizes information for instant recall or context when we need it. In view of these two possibilities, I believe we will have to get much better as time goes on at understanding that we have to balance the growth in available information with the practice of contemplative thinking. Put simply, we need to invest time to digest the information we have imprinted on our brain and build the connections in our minds to turn it into useful knowledge and action. I have spent time with people who specialize in how the conscious and subconscious minds work and they tell me that while we sleep, the subconscious mind continues to work on the information we have gathered during the day in order to put context to it. I certainly have had that experience in my business life when I have been overseas working through translators. I can feel my mind replaying conversations that were hard to understand in real time, all night long. And when I wake up, I have a better handle on what happened the day before. I believe strongly that we will need to learn how to set aside time and develop the habits to absorb the information we gained each day through quiet reflection. If we do not learn how to do this, it is likely we will find that much of the information we take in will go in one ear and out the other (as my mom used to say).

Since I opened the subject of filtering information, we should discuss the good and bad about third parties filtering our information for us. Let's start with the positive. Google has been working for a long time to gather information on who we are and what might be important to us. They gather extensive amounts of information on all of us that use their plethora of free tools. In one way this is fair because their tools generally are free. We get to search for information anytime we want and they never send us a bill. However, there is a downside to using these tools, which is that a company is gathering so much data on each one of us. They use this information for many purposes, including helping advertisers better target us with relevant information. We will talk much more about this later in the book when we take a look at how our privacy levels are changing because of technology. The other way they use our data is to filter our search results so they might be more useful to us. And while this can be helpful, it also is where a troubling dynamic comes to light.

Filtering search results seems helpful on the surface because it provides a better probability that, when we search, we are getting information that is what we want. The dark side of their filtering results is that they are controlling what we see and don't see. For example, if they learn that we are a republican, they would feed us more information that concerns republicans than what a democrat would see because they would assume that is what we are interested in. The problem with this is that then we would get a skewed view of opinions and feedback on politics and, subtly, our beliefs are reinforced and not challenged. This does not lead to enlightened thinking because only through seeing life through many lenses can the Truth be found. If the Internet has the potential to be the fount of all wisdom, Google could be unwittingly crippling our ability to drink from it.

Let's move up the chain to the two levels that are driven by human interaction with data and information: knowledge and wisdom. Data is a raw material that would have little context if viewed in its stored state. Information is collated and organized data so it has more context and can be absorbed more easily. When we jump to knowledge we are stating that the information now has been digested by a person (or is imbedded in software code) so that decisions now can be made by using portions of the information that are pertinent to the decision being made. Then knowledge is enhanced by observing the results of the decisions, which further reinforces the validity of the decision or causes a person to reassess the correctness of the information. A loop begins at this point in the cycle where we use information to make decisions and can validate the accuracy of the information used to make the decision. If the decision is sound and correct, we assume the underlying information was valid. The more information that is memorized or stored, the more knowledgeable a person or system can become.

At each step in the DIKW chain the value rises. Whether we are discussing this from an individual viewpoint or an organizational perspective, there must be a conscious effort to convert information into knowledge so that more intelligent decisions can be made. The more knowledge a person or organization can gather and store (in brains or software systems), the more of a base of intelligence exists to be creative with solutions. This may seem like a basic truth: that the more knowledge someone has, the better decisions and problem-solving capabilities will exist. As simple as that sounds there are many people (and organizations) that do not put a priority on converting information to usable knowledge. We call this process "learning." In a world exploding with data and information, we still have to make a conscious effort to learn. This is all the DIKW concept is: the process of learning and becoming wise.

It is possible to digest huge amounts of information in our minds and then lose it again if we do not commit it to memory and finish the process of learning. This is one advantage computers have since, once information is stored, it is not often forgotten. I believe that we are going to see people get stuck in what I will call the "Information/ Knowledge Spin Cycle" and fail to transfer information into knowledge and wisdom.

The DIKW chain is a consciously progressive process that takes an effort of will to improve. There is not an automatic ability to have each step build on the one before until we get to wisdom all the time. A person can be exposed to data and, if they do not have the will to do the work or apply their brain, they will not move it to useful information. This could be caused by lack of brainpower or laziness. At the next level, information then must be transferred to the brain so that it is held there and it has context of everything else in the brain. Without this it is not really "our" knowledge. Again, this often is by an act of will. Either we choose to learn actively or we learn through experiences that are forced on us.

Here is where I believe generations going forward could struggle. If we have a magic box that can answer almost any question for us and we have decision support systems that can make decisions for us, then why would we work hard to transfer information into knowledge? That takes a lot of work and time; proactive learning takes effort. Why not just become dependent on our "Outboard Brain" to provide knowledge for us? We could potentially skip the knowledge step and just hope to make wise decisions through using our Outboard Brain to help make the decisions for us.

This is already happening. Students in algebra or trigonometry classes can get online and use tools available to answer any advanced math question, yet they would never actually be able to do the math by hand because they know only enough to ask the question of a math application. They do not have the underlying knowledge of the math principles. This is different from the impact of a calculator because just about everyone understands the concept of addition, subtraction or multiplication and could struggle through finding an answer to this type of math problem. They have information, but not the usable knowledge that can be applied, absent the technology. Hence my concern of the possibility of many people falling into the Information/Knowledge Spin Cycle and not progressing is based on this type of phenomenon.

The wonderful new potential of the Internet is that we are just seconds away from retrieving vast amounts of information in order to help us make knowledgeable decisions. We can go back and find information we have forgotten and we can discover completely new information when needed. This gives us the potential to grow our knowledge tremendously for little to no cost. From a very real perspective, a person with an Internet connection can grow their knowledge far past anything that could be learned in a school simply by using free online tools and information sources to grow their knowledge. We just have to avoid the Spin Cycle to get there.

Getting to the level of knowledge and having vast amounts of information in our own brains is worth the work. Knowledge is a powerful asset because it helps us make good decisions, helps us understand options for solving problems and it helps us see opportunities. I am very excited about the fact that young people are converting information to knowledge at a high rate by searching for answers on the Web for any question that comes to their minds. This was not as easy to do in decades past. Today when I survey young people about the percentage of knowledge they get from the Internet versus school or their parents, I am always surprised by how high their answers tend to be. Computers can automate gathering data and turn it into information. As human beings we have to make a conscious choice to turn information into knowledge. It is tragic to live in a time with such a glut of information and yet still have people who are not motivated to gain knowledge.

Experiences play an interesting role in our learning process. In many cases we are provided information (e.g. do not put your hand on the hot stove), but at some point we go against the information provided and we touch the hot stove. This creates a vivid experience that reinforces the information we have been given and we move from information to knowledge in an instant. The Internet provides a base of information that is unlike anything humanity has ever seen. There are trillions of pieces of good information there to be had if we care to do a quick search. For too many people who are not conditioned to use the Web as an information tool, experience will be their best teacher. Experience is a wonderful teacher, however the lessons can be very painful. Over the course of a lifetime we have many chances to gain information, get experience, and create a larger base of knowledge. If this was all there was to the DIKW chain, we could end the story at "knowledge" and just say the goal is to attain as much knowledge as possible. However, in the English language we have words like "wisdom" and "enlightenment," and these seem to have a different and more advanced connotation than just having great knowledge.

We have a distinction we make when we say that someone either is "book smart" or "street smart." What "book smart" means to most of us is that a person has lots of knowledge but is not very good at applying it in practical ways. I mention this only to say that we have made a distinction between believing that having all-book knowledge is the final and best stage of learning. There is something more, which is to attain the highest level of learning or being, and for this we use a word like "wisdom." This connotation expresses that someone not only has knowledge. They have something more; an ability to apply it in wise ways. So what is the basis for a "wise" way? This is the problem with words. We all use them and then, upon further review, the definitions can get complicated. I would like to submit to you two distinctions for the word "wisdom" that make it easy to understand why it is at another level beyond knowledge.

The first is that wisdom is knowledge applied in valuable ways. One can have great knowledge and still not apply it in the practical world with discretion and sense. This is what drove us to distinguish between "book smart" and "street smart." The person on the street actually gets things done in ways that are efficient and with the result in positive outcomes. The book smart person knows a lot but does not always accomplish a lot. It is possible to know a lot about the world and still lack the discretion or ability to use that knowledge to improve outcomes. A brilliant nuclear scientist can make a bomb and allow anyone to control the outcome of its use. They may possess great knowledge and no wisdom. A wise nuclear scientist builds safe and efficient power plants that are a blessing to the world.

This simple example might prompt you to wonder what the magic ingredient is that told that scientist to forego using nuclear fusion as a weapon and, instead, use it as a power source to help the world. Where does this wellspring of wisdom come from? In some cases, it has come from observing how to use knowledge in healthy ways and wanting to continue to use it so that the outcomes are beneficial for the people involved. Yet even this leaves the curious person wanting; where does the motivation to use knowledge in this way come from?

I believe there is a spiritual component to wisdom in that the foundation for wisdom is provided to us through a higher power. We can make rational decisions using knowledge. We can make enlightened decisions with wisdom. I can interact with a person that behaves in negative ways and will observe them doing some activity over and over that is destructive to others around them. This kind of thing can happen on the soccer field while I am coaching a game and observing a referee making a series of terrible calls.

I have a lot of knowledge about the game of soccer because I played for over thirty years and I have coached for twenty years and refereed for years on top of that. I am well-versed in what is legal and what is a penalty. At times referees will make lots of mistakes in a game. I can keep track of the fouls on my team and the other team and gain the knowledge that we have been called for many more infractions in a way that clearly seem unbalanced. From a knowledge standpoint I clearly see there is a problem. Knowledge allows me to have a spirited discussion with the referee and share my observations.

Wisdom stops me from doing so. Wisdom tells me that they are doing the best they can with the ability they have that day. Wisdom tells me that parents and players all would see me challenging the referee and that would be a poor model for everyone watching. Wisdom allows me understand that life is not always fair and that the outcome of the game is not as critical as what my players learn from this competition about how to act in this world. This wisdom is driven by spiritual principals that are at a higher level than cultural norms.

Note to the reader: I have not always been a wise coach. I have been known to use my soccer knowledge while screaming at a referee in the heat of the moment. However, I did pull my team off the field during

111

a game for the championship because the referee was not calling fouls that were injuring our players. I received lots of wisdom points from our parents for putting the girls' health in front of the championship quest. I provide this disclaimer in part so that I am being honest, but also because some of my players might read this book.

Wisdom is not making really good decisions solely because we have loads of earthly knowledge. Wisdom comes from a different place: a universal place of goodness, love and peace. This may be difficult for some people to accept because they see wisdom as being decades of knowledge turning into aged knowledge that is extremely deep and true. While I agree that this can happen, I also see that wisdom cannot be had without adding in the more universal principles built into us through our soul.

This could lead some people to say that they don't care about spirituality or enlightenment so they will seek just to be extremely knowledgeable. This is a choice people can make, but I predict they will suffer from hollow lives in the end because they will be incredibly intelligent abut nowhere near being at peace in this world. The wiser we become, the more well adjusted, joyful, and peaceful we become. This also is a choice we make. If we stall at the level of knowledge and refuse to understand the call of our soul toward gaining wisdom, we will leave much of our potential on the table. Said in another way, God could have created the Internet to help us all be much more knowledgeable. We have to choose specifically to move up to having wisdom by integrating our knowledge with the universal truths of things spiritual; and for this we were given free will.

While you process what might be a very new thought to you about our levels of intellectual maturity, let's get more practical and move on to looking at how the world of the DIKW process is impacting us. We can do that by taking a detailed look at information immersion from the perspective of a typical adult. I am going to break down many different components of information flow and how they impact us. I think this will give you a better sense of how deep the changes are in our world today. Again, I am going to generalize my observations and use trend extrapolation to paint the picture for you.

The overall quality and the volume of information, and how we use it: The Internet is relatively young. We have only had the Web for a couple of decades and, although that might seem like a long time to a young person, it is a flash in time. In just this short ramp up we have already archived a base of information that boggles the mind. Better than just storing it, we have provided a number of abilities to access this information through different interfaces, applications, and devices. We are, at the same time, growing this massive base of information, improving the quality and accuracy of it, and quickening the ability to find what we need. All of this bodes very well for any person or organization that commits to learning: converting this information to useful knowledge. Learning has always been, and still is, a choice. We can be presented with opportunities to learn over and over and still not convert information to knowledge. Or we can search the Web proactively for information every time we are curious or have a need and, thereby, learn something new. We choose how we invest out time; we can binge watch Netflix for eight hours or we could watch for one hour and spend some time learning something valuable.

With that said, what do we search for when we go online? According to Google (And this is curated by them to take out anything having to do with sex or porn) here is what people searched for in 2015:

Top Searches by Adults According to Google in 2015

1. Lamar Odom
2. Jurassic World
3. American Sniper
4. Caitlyn Jenner
5. Ronda Rousey
6. Paris
7. Agario
8. Chris Kyle
9. Fallout 4
10. Straight Outta Compton

As you can see above, the subject range is pretty vast and tends to lean towards the hot topics of the moment. By the time you read this you might be confused as to why some of these search terms were in the top ten! The impact on us of being able to access any piece of needed information instantly gives us the power to self-teach in a way that would have been difficult earlier in our lives.

At work we have the ability to be much more productive because we can access huge amounts of information in order to speed up whatever task we might be assigned. The odds that we can find accurate information in order to make better decisions are higher now. The need to "guess" at answers is lessened every day as we can depend much more on fact-based decision making. We have much more ability to look around at what others are doing in order to speed up innovation. We know what our competitors are doing much earlier, so the pace of change in the market just continues to speed up. Transactions are happening faster and easier now and the number of companies we buy from has grown because we are not restricted to buying from local vendors due to physical geography. Suffice it to say that the Internet has transformed many parts of our lives from the moment we wake up until we go to bed.

At home we have the ability to buy from anywhere without leaving the room. We have access to any form of medical information at our fingertips. We can look up how to solve any problem that may arise, from fixing a broken washing machine to cooking a specific meal. We can communicate with hundreds of friends and relatives virtually for free and instantly and hence, have better visibility into what is going on with people we choose to follow than any other time in history.

We can curate any type of information, picture, or video that is of interest to us so that we have huge libraries of content we can use for work or enjoyment. We can spend hours discovering this information online and storing it locally or in the cloud. Curation of information helps make us experts in the categories on which we focus.

As much as we talk about the younger generations being heavy technology users, when it comes to information usage, adults probably spend more time online than kids. Adding knowledge is so important in adult life because it defines how far we go in our careers, how well we do as parents, and how successfully we navigate the grownup world of responsibilities.

Real world impacts: Looking at this over time, if a person chose not to search the Web when they had a question or a thought, we can see how, throughout a life, there would be hundreds of thousands of pieces of information that would never get into their mind. Looking at it mathematically, if an adult lives for 20,000 days and in each day there are ten pieces of information they did not get that crossed their mind, this alone is 200,000 pieces of information never gained. Now for those of you that think ten pieces of new information is a lot, compare this to what we have today. Not only do we search multiple times a day; we read streams of information on social sites and other informational Websites everyday. I would guess that we digest hundreds of pieces of information we would not have in the

114

past. It's likely we are talking about millions of pieces of information we have today that we would not have had 25 years ago.

The more information and knowledge we can harvest; the better decisions we make. The better decisions we make, the more enlightened we become. The more enlightened we become, the better model we are for those around us.

Positives: Clearly having access to information on demand is something that is changing the world in many positive ways. To be honest, another whole book could be written just to identify major examples. Having good information on any subject illuminates the truth for us, and the truth is a great thing. Many a poor decision has been made in history simply due to a lack of correct information. Many people have died because they or someone else misunderstood the truth. I really believe that the single greatest asset of technology today is the dramatic improvement of our ability to share the truth with each other from any corner of the world, instantly.

For example, many dictators have (and still are today) committing heinous crimes against humanity. If all the people of the world could see, unedited, what was being done, these regimes might have been ousted more quickly. Today no dictator can hide what he or she is doing successfully because citizen journalists get online reports out somehow. At the same time, there are many countries that believe now it is their duty to protect the innocent and downtrodden. At a macro level this is a trend that has far-reaching impacts on the human race. Instead of isolated pockets of terror, rape and death around the globe that act in darkness, we have technology shining a light and countries banding together to help.

Another example of how our growing information base is changing our lives is in the medical field. Even in our modern era, many health problems have gone misdiagnosed or undiagnosed because of a lack of information that could have resulted in an accurate diagnosis. The Internet gives people a fighting chance at least to explore possible causes or remedies on their own to supplement a practitioner's expertise, or even to replace them. This certainly is a better situation than getting one person's opinion with no other check-and-balance. I do not mean to offend doctors, however even with all our modern diagnostic equipment, they are, more often than not, guessing when diagnosing many maladies. I acknowledge also that health care professionals do wonderful things for many people by delivering correct treatments. Until collectively we can diagnose a health problem with 100% accuracy we need to get real about the positive potential of the Internet in providing healthcare information to people directly. It is logical to understand that, in an era when many doctors are limited to a fifteen-minute engagement with a patient and are diagnosing from just a few physical pieces of data, there are going to be failures in the system quite often.

I almost want to avoid telling this story, however it happened and is a good example of the life-altering power of having access to information at the touch of a button. My wife Annette had spent a year dealing with a slowly-worsening physical problem. She was struggling when she ate, lacked energy and was having back pain that did not seem to emanate from a spine issue. It came and went but never got better. It was getting worse slowly and was to a point that it had degraded her quality of life seriously. It pained me to see her like this because normally she is a really happy and cheerful person, but it was clear she was struggling with a lot of pain at times. After going to seven different types of doctors in different cities, she was still struggling with the symptoms and not getting better. One day while lying on the couch at home, I decided I might be able to help just by using logic to find some other possibilities that she could check out. I had heard about the wife of a friend who had her gall bladder removed and her mysterious symptoms disappeared. Armed with that information I searched on the Internet for information about gall bladders and their ailments. Then I shared with Annette that it was possible this was her problem. She went promptly to an emergency room where she had to convince a doctor to do an MRI (he did not see the reason for this and she explained that she would pay for it and wanted it done to check her gall bladder). When the results came back, she had a pretty serious case of gallstones. She had it taken out and, once she recovered from the surgery, was pretty much back to normal. Most likely I would not have gone to a library and checked out a medical book in the past, but in this case having the instant access to a huge amount of information on gall bladders made a difference for Annette.

Stepping back, if we look at information flow and how people form viewpoints on the world around them, it is clear that a wide flow of opinions and data can help a person find the truth more clearly. When people are exposed to a limited number of viewpoints in life there is a chance they will be narrow-minded and, in some cases, horribly misled about what is healthy or unhealthy in life. Generations can pass down their biases and bigotries because there was no ability to self-learn about other viewpoints. Today there really is no excuse for not absorbing a well-rounded view of the world. This goes for citizens of countries whose leaders try to mislead about the truth or sons and daughters of bigoted, hate-filled parents who try to pass on their psychosis to their offspring. The Internet provides instant access to viewpoints from all over the world and, when taken as a whole, these will balance people out in a healthy way.

These are just a few examples of how a high information flow helps the world. A person's ability to digest a large river of information always will result in them becoming more enlightened in life because, over time, they will be able to discern the Truth by assembling information in order to have enough accurate visibility that they can actually discern truth from fiction. Ignorance lives in darkness and we have always known that.

Negatives: When I use the word "information" it sounds pretty benign on the surface, and if it was just a collection of pieces of data like where to eat or the score of the last basketball game, there really would be no negatives. However, when we talk about all of the information on the Web, we have to cover every video, every picture and every comment made by anyone online. This means that any unhealthy behavior that an adult might engage in can be supplied or enriched through content available on any device, any time. Of course the list of unhealthy activities is long, so let's just focus on a few.

Information about people we have had past relationships with has become a new dilemma on the Web. On one hand we have "revenge porn," which is the practice of people posting nude pictures of the person they have broken up with. On the other hand, we have people digging up past relationships and breathing life into them because connecting with anyone is so easy these days. Whereas in the past we might lose touch forever with an old boyfriend or girlfriend, today we can look them up and connect in minutes.

Another form of information that has exploded is porn and, while this has been a problem since the days we could first take pictures and shoot movies, it has become more pervasive since the Web made access cheap and easy. In fact the Web even offered new options for interacting with the objects of lust online.

Porn Sites Get More Visitors Each Month Than Netflix, Amazon and Twitter Combined.

30% of the Internet industry is pornography.

Mobile porn is expected to reach $2.8 billion by 2015.

The United States is the largest producer and exporter of hard core pornographic DVDs and web material, followed by Germany.

Source: internetsafety101.org

There are those who will argue that porn is perfectly healthy for a grown adult to indulge in if used in the right ways. I must beg to differ because it can cause damage to the psyche of the people creating it and it does little that is positive for those who consume it. I suppose there always will be people who will argue that everything in moderation can be healthy for us: drugs, alcohol, porn, etc. But I don't agree. For the one user who is not damaged, there are many who are.

A completely different negative is the overwhelming volume of information coming at us through our devices. Accessing it takes time away from other activities like relationships, rest, and fun, while not accessing it can leave us uninformed on critical happenings. Until we get better software to help us filter all the information we would love to digest, often we are forced just to grab bits and pieces as it flows by us. This is leaving some people with a helpless feeling because they see no way to harvest and digest all that they would like to turn into knowledge.

Another negative of our giant information flow is the distraction factor that real-time delivery brings with it. Having access to an expansive and never-ending river of information can create an addiction to accessing it, especially if you love to learn as much as I do. Not only is there a huge amount of information available to us; there is a lot of great information that is really fun to learn too. The flow is just about endless, which is where the problem comes in. If we love to gain knowledge, we can literally spend most of our day accessing information and still not be satisfied. The result, of course, is that we will have ignored lots of other parts of our life. This can be seen with people who are scrolling endlessly through Facebook and Instagram, or business-people endlessly trolling through news sites in order to stay up on the latest information.

Without realizing it, one can fall into the trap of standing in their river of information until they drown out some of the more important aspects of life like their relationships with their spouse and kids. They can even compromise their long-term health by becoming more complacent and not moving around or forgoing meaningful exercise. Sucking in information can be seductive because some of us have a driving need to learn the Truth of things.

The speed of access: Adults have different needs than children when it comes to the speed of information flow. We live in a world where parents, employees, and senior citizens all are under pressure to solve problems, make decisions, and help those around them. That means any tool that can provide instant information (answers) or give us a path to solving a problem is going to get heavy use. At work we need information to make decisions, and the faster is better. At home we need information to solve problems so we can get things off of our list. In both worlds we want information to help us learn so that we have the knowledge to accomplish our goals in life.

There is an assumption already that a person will be able to find an answer to any question instantly. We are frustrated if we want to know a thing and it takes even a minute to find it. We expect instant information gratification. Think about how frustrating it is to go to an organization's Website and not be able to find some basic piece of information like a telephone number on a page. Have you ever tried to look up an old friend and found they were invisible online? It just seems strange to us now that, within moments of thinking of someone from the past, we cannot get pages of information on him or her. What about basics like the time when a movie will start, or the hours that a restaurant is open? It really does not matter what kind of information we want at this point; we expect it to be at our fingertips from any device. It is interesting to me that this greed for information speed happened so fast because we did not have this expectation even fifteen years ago. Now "searching" online has become a generic activity in which we expect answers to every question.

Why do so many people love text messaging to the degree that they send thousands each month? It is fast and convenient. It is the most efficient method today for exchanging information or asking a question. We could do anything we do through texting by email or with a phone call for that matter, but we have learned that the responses from texting come much faster. Most people stop to look at a text when it comes in and most will answer a question asked right away. We can access a text while doing about anything else, and we do. We can send texts while doing about anything else, and we do. It is all about the speed of sharing information with this medium.

We have become more of an instant-gratification society and an example of that came when we invented the mobile phone so we could exchange verbal information instantly instead of waiting for someone to check the answering machine. In our world of devices, the desire to access information quickly is now driving the wearable paradigm where we use devices that are always on our bodies so we are instantly alerted to a contact or information. By putting technology on our person, we are again speeding up access time. When Internet-delivered information is displayed in our field of vision or spoken to us through a "hearable" device, we get information even faster. We are just beginning to see the types of devices that will speed up our ability to access information by search or from a river of information pushed to us. We are impatient with information access times already and we have only had these capabilities for a handful of years.

Real-world impacts: There are many ramifications of speeding up access to information and, surprisingly, they are balanced between positive and negative. The very fact that we are working so hard to continue to speed up the timeliness of information and the ability to consume it seconds faster (through wearable devices) shows us that we have a greed for getting all the information we want immediately as it happens, or to retrieve a solution in an instant. All of this impacts our quality of life, how well we can solve problems, how smart we get, how fast we get smarter and ultimately changes the course of our lives many times a day.

Because we go to the Internet for information, we depend less on humans around us and less on books. This changes human-to-human relationships and our ability to read for extended sessions, not including reading for entertainment. The quality and breadth of the information we access online helps us make fewer mistakes because we do not have to learn our lessons from poor sources. Of course there is incorrect information online and when we do a search we have many options instantly for the same information so that we can do quick comparisons.

For a mobile device-carrying person who is connected to social sites, and well-versed in how to search for the information they want, there will be dozens of moments a day when they will interact with their device and, based on the information gleaned, take a different action than they would have taken in years past. Add up those dozens of times a day and you have a huge impact on our lives through information immersion.

Positives: I have already mentioned some of them in the narrative so far, so allow me to be concise in rounding up a list:

The more information we can access with a search, the more likely we will be able to discern what is correct and helpful to us. The faster we can access the information, the more valuable it can be in solving problems. This is not a small benefit because we face solving many problems every day. Having tools to help us solve problems faster and better improves many aspects of our lives.

Having a real-time flow of information about those around us allows us to help someone we care about who is in need. Whereas we might not have learned about their problem, or learned about it days or weeks after it happened, today we can see a need and get help to them instantly.

The more we have options to learn about anything we want online, the less we are stuck with being influenced by a small group of people around us. This will, over time, help us grow out of the bigotry and narrow-mindedness that could come from adults passing on their issues to their children.

Because we had less information in the past, we tended to make our decisions from instinct and experience and not through fact-based decision making. This led to many poor decisions. Today we have an increasing ability to learn the facts before making decisions so that, by applying fact-based decision making, we improve the chances of making correct decisions.

The more humanity archives all our knowledge in a freely available resource (the Web), the more we have the ability to stand on the shoulders of those who went before us so we can launch from what they learned and expand our collective knowledge base.

Negatives: In the business world, we talk about data being a raw material or an asset. As such, it can be used for good or bad. Information is no different in that it is a raw material that we apply in our lives, and there are as many negative results as positive ones.

With worldwide free-flowing information that is stored online for years, we have introduced a problem with regard to our online reputations. One branch of information is the data that resides online about each of us. For a growing number of people, we are talking about years and years of information and thousands of pictures, comments, videos and opinions. If we have a positive reputation, all is good. However, the dark side of having lots of information available when someone searches for our name is the possibility of someone running a "badvocacy" campaign on us, or that something from our past may haunt us forever. In either case, this information may unfairly paint an inaccurate picture of who we are today for employers or possible friends.

For many people, the ease of searching for information has made them sloppy at having the discretion for making sure they are finding accurate information. It is too easy to link to the first online source a person finds, just assuming it is correct. Because we are under more time pressure with information retrieval, we also are less motivated to vet the information.

The speed at which information travels now allows misinformation to spread to millions of people before it is debunked. In the old days when information spread slowly, there was plenty of time to verify it and stop passing it on if not correct. Today with tools like Twitter, a false truth can be posted and re-tweeted to millions of people within minutes. There is not the friction in the system with editorial oversight to slow down bad information any more than good information.

With the convenience of information access also comes dependence. Our devices become more outboard brains that augment what our real brains do. As we make this integration we become more dependent upon - and addicted to - our devices and their connection to the Universal Mind. This creates two dangers: the anxiety that comes with being disconnected, and the possible weakness we end up having when not connected. For example, as the Internet becomes the one-stop shop for solving most of our problems, we lose some of our native problem-solving skill. We are not exercising that muscle as much, so when we need it and don't have technology to augment us, we are weakened.

The last negative to consider is the danger of losing our individuality through our connection to information flowing from the Universal Mind. Sometimes we use the words "group think" to describe what happens when a group of people all convince themselves of the same set of "facts" and combine this with the desire to be accepted by the group. This homogenizes thought and behavior and, in some cases, ruins creativity and innovation. With information now so globally available, there is a danger that when someone searches and the results are filtered by a company like Google, we step closer to group think. When we all were disconnected from each other electronically, it was easier for a person to develop their own interpretations of a subject.

Just to close out these thoughts about how information access is changing us, let's consider the ultimate in speed for getting information: having it pushed to us before we know we need it. There are many examples of companies trying to provide solutions that bring information to people through predicting what they might need before they are aware they might want it. This crosses the line between fast access to information and into predictive delivery of information. A car that tells its driver that they may be getting sleepy or that someone in front of them might be getting ready for a fast stop uses such predictive delivery. In a healthcare sense, the devices we wear gathering heath-data and warning us in advance that we might have a medical situation arising soon anticipate issues we don't have yet but may down the line.

Let's finish up examining the concept of the DIKW by looking at where we are likely to go in the future. The following is a list of specific concepts I believe we will see going forward:

The volume increases, and the speed at which we can access it improves: This is an easy prediction because we have been headed along this curve for a while. Still it must be mentioned that we are still in the early stages of creating, harvesting and archiving information as a process. The majority of information generated today is done so by regular people, not organizations. As we move toward the increasing ubiquity of the Internet of Things among us and have billions of devices joining us in creating and flowing information, our ability to know what is going on around us will blossom. At the same time, we will continue to develop data structures and devices that will flow the information in real time for our consumption. This ability to know everything going on around us will seem a birthright.

We will get new sophisticated tools for filtering information: The best way to improve the quality of the information we can glean from the Web is to be able to filter it through a series of conditions that weed out what we don't want so that we can digest what we do want quickly. Soon we will have information-harvesting applications that will allow us to provide a whole suite of filters so that we get exactly what we want, when we want it, and on the best device to consume it from. These filters will be configured into customized combinations so that a vast array of information from people, the things we own and the groups we are affiliated with will synchronize in our minds effortlessly. If I want to see all of the soccer goals scored by a specific team in the Premier League, I would assemble filters that would only monitor those games my team played and automatically would grab a minute before and after the goal. Then it would push this to my mobile device and give me an alert so I get a text message with a link to the video of the goal. I could even add an archive filter so that all of the goals would be compiled into a yearly video and stored in the cloud for me to watch anytime I choose.

Information assistants that search on our behalf and bring us the information we want: We will have an online avatar that knows us and acts as our personal librarian. These pieces of software code will watch everything we search for, learn about what interests us, and see what we need in order to do our jobs. The avatar will use predictive analysis to search the Web constantly for every possible piece of information it thinks we might want or need. It will be visually represented as a person or in whatever form we choose, with the type of voice and tone we like to hear. The avatar will be portable across devices and will appear to us when we call it by voice. This will work great because we have unique voice patterns, so by combining our voice with a key phrase, our information assistant always will be ready to pop on any screen the moment we call for it. We will have a powerful ability to modify our avatars so they weave into our lives in exactly the way we need for them to show up. Over our lives they will get to know us better and we will grow quite attached to them because they will have solved thousands of problems for us over time. An entire industry will grow around providing these avatars and then backing up the many years' worth of data they will have learned about our needs and desires. This will be a critical service and we will pay for it gladly because it will be the foundation of the avatar's ability to serve us.

Sophisticated personal listening/alert systems: Whether we use an avatar to listen and alert us to desired information or not, we will have a powerful ability to design information alert systems that notify us whenever a new piece of information that fits our parameters hits the Web. We have social listening systems today that are used by companies to monitor anything customers say about their brands or products online, so we are building early forms of these alert systems already. Soon the power of these systems will move to individual use and we will use them less to listen for information about us being posted online and more to alert us when pieces of information about things we care about are posted. Whether this is the score of a game from our sports teams, treatments for multiple sclerosis, news on specific friends, or any video posted of a wiener dog, we will be able to create a template of all the information we want to absorb, route it to any of our devices and access it through any of our delivery channels. This will improve our current river of information so that we can absorb more of what we care about in real time.

Simple storage and retrieval systems: As we improve the ability to access information that we care about, we will need better ways to store what is important to us so we can access it later. The tools we have today are first-generation and pretty crude. Going forward, with the simple wave of a hand or verbal command, we will be able to store information into whatever framework we have designed to hold it. Depending on which device we are using to consume the information, we will have many options for routing the information. A wearable device will let us read an alert and simply by closing our fist will archive it intelligently with our storage system, guessing at where we want it saved. If we are driving and see information in a heads-up display, we will ask verbally for it to be stored and, once again, the system will make an assumption for us about where to put it. All of this information will be stored in the cloud so we won't worry about it being available on a specific device or about it being lost when one of our devices breaks or is stolen. When we want to retrieve our personal information, we will ask for it verbally or with a few typed keywords and it will be found instantly. Access to a lifetime of archived information will be available when and where we want it.

Timeframe sliders that show forensic, real-time and predictive states of information: Since all data and information has these three time states, it would make sense for us to have the capability to view information with these dimensions. If you think this is a crazy idea, note that social sites online are providing your data in timeline formats already. An example: If I wanted to learn about a company, I could bring up their profile online and then use a slider at the top of the screen to scroll back through all of the information on the company backwards to the day it was founded. If I want to see what they did in a specific year, I could scroll back to that. Then I could scroll forward to what they are doing right at this moment and even cross over to any type of forward-looking predictions being made about the company's future plans. Today most of the information we absorb is forensic in nature and this will change as we move more toward valuing real-time and predictive states of data. Think about applying this new capability to scroll in multiple time states to your child's school, a sports team's performance or a product you are interested in purchasing.

Easy data/information mash-up capabilities: One of the more enlightening things we have learned to do with data is to "mash up" multiple sources so that we can see trends or patterns that would not be clear by looking at the sources individually. So if we took the customer transaction information for a company and overlaid that with Google Maps, we might see a heat map that is color-coded for where the customers live. This could help us understand how far people are willing to drive or where we have neighborhoods that did not connect with the company. Putting together mashups is a growing art form done by data scientists because it can be quite illuminating. As so often happens, a capability that is built for companies will migrate to us as individuals. We will have lots of tools that will help us access thousands of information sources and we will combine them in many different ways in order to gain insights we could not get otherwise. Just in case you are thinking this is something you would never do, please understand the tools we will use will make it very simple to do things like overlay the travel patterns of a child with Google Maps so we can see every house or store they have visited in the past six months. We will think nothing of being creative in combining sources of information in recipes we cannot conceive of today.

Simpler information visualization capabilities: A critical area of improvement we all seek is to have better and faster ways to digest information. When we say a "picture is worth a thousand words," we are proving this point because the graphical representation of something can tell us more in a glance than reading about it. You can tell me in 800 words the horror of a murder case or you can just show me a picture of the body and the murderer. Video provides an even more intense method for delivering the information because it provides thousands of data points flowing across our eyes. According to Forrester Research, one minute of video is worth 1.8 million words! We absorb information so visualization is key because it provides a method to do this absorption much better than digesting words and numbers in a raw form. In the future we will have much better ways to search for information by using visual interfaces instead of a search box and these interfaces will provide a much faster and more accurate ability to get to the exact piece of information we want. You see the early seeds in our ability today to choose a picture and have the search engine bring back every other picture that looks like the one chosen. Now imagine what happens when we have an ability to do a 3-D holographic search with gesture control like we have seen in a number of science fiction movies. There is a reason that science fiction writers have been describing visual search for years now; it will be a much better way to find the information we want. At the same time we will take much of what we are learning in the field of data visualization and apply it to information visualization. So when we ask for the scores of an NFL team we will have a 3-D model that is scrollable with a time axis and a detail axis so we can get to any trend or individual insight we are looking for. We will depend less and less on black text on a white screen to deliver information and we will move toward rich media sources that can tell us the story faster and better.

Information brokers: Information on the Internet today is most often free with only a small number of organizations actually charging for content. I believe that we will move slowly to a model where we will pay micro-cents for various types of information. This will include real-time and predictive information because this will have more value to us and is harder to deliver. This will include aggregated and visualized information also that has more value because it will be pre-formatted to help us absorb it quicker. As we get better at watermarking digital information and improve the rights management, it will get easier to assign value and restrict the flow of information. This has happened when you think about the MP3 format for music and the transition from Napster to iTunes. In the beginning music went digital and was traded between people for free with no ability to control the distribution by the artists or labels. Then iTunes came along and improved the quality of the file, restricted the delivery, charged a small fee and many people switched from consuming music for free to a pay model. While there still will be a huge amount of information that is free there will be certain types that will be "sold" by information brokers who control the distribution and they will monitor the sharing as well. An example of this might be charging fans to access the digital life logging information of one of their favorite stars. Another would be charging people to access deep product information on every good provided at stores today. We have precedent for this already with sites like Angie's List which charges people for digital information about service providers.

One additional note when talking about information brokers and what they will become: I am sure we will see a growing black market for information that people should not have access to which will be sold illicitly. We see the early seeds of this as hackers who steal profile and financial information sell them on the Dark Web. The logical next step for the criminal-minded will be to capture all kinds of private information (for example, healthcare histories) and sell it to anyone who wants to traffic it online. The financial rewards of selling data illicitly will then lead to proprietary business data bases being stolen and sold on black markets. If data can be sold out of the back of a truck in a dark alley, someone would do it, yet there is not a need for this when we have the Dark Web and Bitcoin because the combination of these provides all the anonymity that criminals need.

Information is a bit like the air we breathe; we take it for granted until we don't have what we need and are left in a lurch. When we have plenty of it we cease to recognize how often we use it and what it means to us. As our ability to access it grows more sophisticated and as the volume available increases, we will find more ways for information to impact our lives, both positively and negatively.

As I have said we are likely growing numb to the fact that information volumes, quality, and availability are dramatically changing us and how we live our lives.

From both the section above and your own experience, it is pretty clear that we are in a completely different situation now with the huge amounts of instant information we will have access to from now on.

The bottom line is that endless access to information does not necessarily translate to wisdom. We only attain wisdom by learning to apply knowledge in ways that are healthy for humanity and, ultimately, that are driven by spiritual principles like goodness, forgiveness, grace, and kindness. The Internet cannot give us wisdom; it can only provide us information that we must migrate to knowledge and then use to make wise decisions. We would benefit as a species to be conscious of this process and make proactive decisions about how we view the importance of moving data from information to knowledge.

We have been on a march for the past two decades toward moving information from paper to digital media. Along the way we began to move all the information we possibly can out of our minds too and onto the Web through blogs, videos, and billions of observations and snippets on social sites. This means we are creating an accessible reservoir of a much higher percentage of our collective knowledge. As we then improve our ability to filter, search, and flow that information in real time, we will change thousands of human dynamics and billions of decisions each day. There is no overstating the impact of the improvements we are making with immersing ourselves in data, moving it to information, converting it to useful knowledge and finally – hopefully – using it wisely.

SUMMARY:
Information Immersion

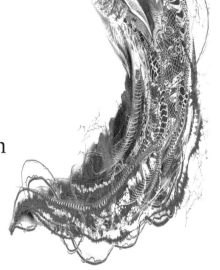

We are improving our ability to harvest data, store it, and actually turn much of it into useful information quite dramatically. We have the potential to use our computers and mobile devices to access all of this information and turn it into knowledge we hold in our own minds. What remains to be seen is how well we can find the Truth within all our information and apply it to our lives and decision making so that the vast enhancement of our capabilities actually does more for us than find the best Chinese restaurant close by or fill us in about what our friends are doing at the moment.

Our growing ability to accumulate and dispense vast amounts of information does not necessarily create wisdom. Wisdom comes from converting information into knowledge and using that knowledge to discern the Truth by then integrating spiritual principles. A society becomes more enlightened, healthy and progressive by growing the percentage of its people who make good decisions based on wise choices. This all sounds reasonable and simple when stated this way, however in the real world, we have millions of people who make terrible decisions every day because they either lack good information or they have failed to grow knowledge into wisdom.

I believe there is no form of spiritual belief where a desire to be ignorant is valued. In order to grow and progress, humanity must continue to improve our ability to learn. The more intelligent we become, the more enlightened we can become. That is exactly where we will go with the next chapter.

My light-vs.-dark ratio on information immersion is 80/20 with a heavy advantage on the positive side. Although we are exposing ourselves to information/content that is not healthy sometimes, for the most part we are trafficking in information that is helpful and positive. The technology impact is in providing wonderful potential for us to learn. All we need do is put in the effort.

Humalogy Viewpoint

When looking at our processes for moving data to wisdom in the past compared with our abilities today, I believe we have moved to a current score of T1.

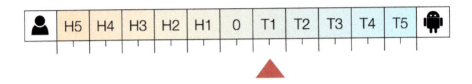

What I see today is that we have a much better ability to gather and store data than to get value out of it. That is not a problem with the technology; it is a lack of human skill in being able to mine the data and turn it into useful knowledge. We have much better tools for providing the ability to move up the DIKW chain than we have people with the skills and talent to use the tools. As time goes on I hope to see the Humalogy score shift toward the "H" side a couple of notches as we get more sophisticated in using the information we are generating now in our Big Data world.

Seminal Questions

What can we do to teach our young people how best to transform information into wisdom and then to have the will to learn and improve their own minds without becoming overly dependent on technology for solutions?

Ultimately will a powerful flow of information make us organically smarter or simply more dependent on technology in order to be intelligent? Will it make it too easy someday for many of us to be stuck in the information/knowledge spin cycle because we depend on our devices to provide temporary information/knowledge we fail to retain?

How much faster might society progress now that we are standing on the shoulders of past generations by archiving vast amounts of data and by being able to filter and manipulate it instantly?

How far might we go with integrating digital tools into our lives and our bodies in order to access information when we want it and how we want it? Would we become like cyborgs in order to have instant access to every bit of information available?

How can we best help each other understand the critical importance of the difference between making a culturally knowledgeable decision and a spiritually wise decision?

CHAPTER FIVE:
Digital Learning

"The illiterate of the 21st century will not be those who cannot read and write, but those who cannot learn, unlearn, and relearn."

—Alvin Toffler

In the last chapter we touched on the difference between consuming information (that flows into our minds but does not really land there) and converting it into knowledge. We choose thousands of times a day to learn something or move on down the road without getting smarter. I made the case that the DIKW chain is a model that describes how we increase the value of data and, in doing so, we learn. In this chapter I want to deal with how technology influences our capacity to add valuable knowledge and why we will do this at ever-increasing rates. Most societies in our world understand and value learning although there are a few unenlightened places where leaders still try to restrict education as a means of control. This demonstrates how far away we are still from being enlightened as a species.

Learning is the process of moving information to knowledge. Throughout human history we have sought to improve the tools we use to get this done. I suspect that our passion for learning has gone through many eras of different iterations. For many thousands of years learning likely was valued for survival reasons. As our early ancestors learned to use crude tools and found processes that could make life easier and more painless, the odds of survival were enhanced. The more an older generation could teach a younger group, the more survivability was enhanced because each succeeding generation could build on the lessons learned from the ancestors. This has been true all the way to our present time. This is the reason our average life span keeps increasing.

Somewhere along the way humanity was surviving well enough to look around and gain a curiosity for the things around them. We began to wonder about the stars, about the weather, about the cycles of time, gravity and how raw materials could be combined to create new things. Thus began a hunger for knowledge that impacted our survivability only partially. In many ways this was a hunger to understand and learn about the world around us. This caused learning to flourish in order to explain every physical thing around us.

Then we moved on to curiosity about ourselves: Why do we exist, why were we put here, what were we designed to do and what does it take for us to be happy? This ushered in an age of philosophy. For centuries now we have endeavored to answer these existential questions so that we could learn more about humanity, the reasons for our existence and the keys to more than just surviving; we want love, joy and peace. We are still in this era, still trying to respond satisfactorily to these very deep and important questions. We have learned that not having answers to these questions can leave a person adrift in life without a set of beliefs to act as an anchor.

Our hunger for learning has served us well. We have become more civilized in each successive generation. Looked at collectively around the world, we are still far from being completely civilized. We see acts of brutish violence everyday somewhere in the world. With that said, we are much more civilized today than we were even three generations ago if compared against the horror that what was done during World War II by the Nazi regime. It is difficult to step back and see how we have grown over the scope of thousands of years because we only experience our own time, with the actions of our ancestors relegated to history books. I believe we are becoming more civilized and enlightened steadily; we have our hunger for learning to thank for that. This is the reason we must examine the role that technology is playing in helping us learn. If digital tools can accelerate our ability to learn greatly, we may also accelerate the rate at which we become more enlightened. That is a very hopeful thought.

Some of you might be thinking this chapter will be heavy with comments about how schools need to rebuild how they teach, and there will be a bit of that. Actually this is much more about the lifelong impact of technology on our learning ecosystems, which is much more important in the big scheme of things than what we do in school as young people. I say this because we spend much more of our lives learning outside of a learning institution. We are adding every bit as much knowledge as adults than we did as children. Sometimes we hold this knowledge for our lifetime; in some cases we hold it for minutes. While I know that, every day, information goes in and back out of our brains, I also know that most of us value holding a good amount of our hard-earned knowledge permanently.

When I was younger, I would hear people say, "Knowledge is power." This made sense to me. I would hear from elders, "If a person has vast knowledge, this is something that can never be taken away. You can strip a person of everything they have, drop them in the middle of New York City and they would be able to build back everything they had if they just have deep knowledge of the world." I suppose this inspired me to become an avid reader and to seek to learn as many new things as possible. You never know when you might be dropped naked into NYC and have to survive, of course. With that said, I was not a great student in school and never went to college. I do, however, spend a lot of time every day reading and learning, and have done this since I was young. At some point I must have caught up to the people who went to college and got good grades.

I mention this just to point out that lifelong learning has very little to do with what school you went to or what kind of grades you were given. Over the course of one's life we can acquire incredible amounts of knowledge regardless of our early education. In fact there are billions of people around the world who do not have access to excellent schools and sixteen years of formal education. For these people technology is a great leveler because it provides access to all of the same information that any high-end private school provides. This is no small new dynamic and, as time goes on, the availability of free online learning will continue to grow and improve. This means every person with Internet access will have the ability to build a learning ecosystem to improve their status in the world.

Getting back to knowledge being power, let me make two important points about where we are today with knowledge. First, there is an ultimate Truth. Of course there are some situations in the world that defy right-or-wrong answers (should I take this job or that job?) but there are many truths in the world that absolutely fall within a black-and-white category. There also are many things in the world where the Truth is determined by what is healthy or not healthy and not so much right or wrong (what should I eat for lunch this afternoon?). Regardless of how we feel about God or spirituality, it makes sense that there is a Truth about some things in our world. Gaining knowledge is a journey toward the Truth. Sometimes we use the word "enlightenment" for the practice of gaining knowledge because it uses the metaphor of having more light or being able to understand more clearly and deeply. For this reason the process of learning is fundamental because our ability to learn is what allows us to gain knowledge and seek the Truth. Learning is a choice and many people choose to stay ignorant. For them technology will make little difference. This is a tragedy at a human level not only for that person, but for all of us who have to deal with them as well.

Second, we live in a knowledge economy now, which means that the "smarter" we are, the more likely we are to prosper. Harvard did an interesting study where they tried to put a dollar amount on what one IQ point means financially. What they were able to show is that if we compare two siblings and one can get a one point IQ advantage on the other, it equates to just over $800 a year in earnings. While this might not seem like much it is important evidence that the smarter people have an advantage in this economy. So if we accept that we are in a knowledge economy and that the smarter people will prosper, we have to ask ourselves why people don't work harder to build a learning ecosystem and move up the curve, especially when we have wonderful technology tools that cost very little to help them. After all, we do not live in a time where information is hidden from us, hard to reach, or too expensive to obtain.

We live in a world where the opposite is true now; this includes people all over the world in all kinds of economic conditions. Not every person in the world can afford a mobile device with Internet access but the numbers who can are growing, and technology costs are dropping consistently. This means that most people have the tools and ability to add huge amounts of knowledge if they will dedicate themselves to technology-driven learning. This is not a small benefit because it allows a person who would not have had the chance to learn at a high level traditionally to be able to build much the same kind of learning ecosystem as the wealthiest people.

Finally, knowledge can equate to power, yet it is Truth that is found in wisdom. Power and knowledge are cultural level states that are used to have a great ability to get things done or have great control. Truth and Wisdom are a higher-level state where the goals are not to exert control or get things done. Truth and Wisdom describe an enlightened way of being that seeks to connect, inform and flow love to be helpful. For this reason we need to be careful about how we interpret the phrase "knowledge is power." If all you seek are cultural levels of power being incredibly smart might get you there. If you seek a higher level of existence, wisdom is the state you seek; this might reverse any desire to have power over at the cultural level.

Regardless of the level you seek to attain in life, learning is the magic wand to get you there. Ignorant people do not become successful even when they win the lottery, mostly because their ignorance can make all the money in the world go away. Life-long learning is the path of a successful and well adjusted person. We are still living with a hangover from a time when we believed that "learning" was done in schools or training classes, and that the vast majority was done when we were younger. Many also have a lingering viewpoint that learning is done through an institution and if we want "real" knowledge we have to get it through a school, a university, or as an apprentice to be taught by a mentor one-on-one. This is not true anymore.

Anyone with an Internet connection and a device now can learn whatever they want, whenever they want and, in many cases, from whatever source they choose. Granted, there are some skills that only can be learned by practicing a task over and over physically, so I don't want to get carried away with what can be learned by reading or watching a screen alone. With that said, the ability to access such a vast base of information the instant we are curious about anything provides a powerful ability to self-learn at a degree far past anything mankind has had to this point. Our new reality is that we have tools that allow us to be inquisitive about any piece of information, do a search and turn it into knowledge in seconds.

People who use, but don't build, digital tools rarely give much thought to what's behind our ability to acquire knowledge online, so here is a quick primer of the tools that most people take for granted:

- The Internet itself provides instant access to the collected, archived, sorted and edited base of a large percentage of human knowledge – all in one place, so to speak. It provides the standards and protocols to keep this flow of information somewhat disciplined as well. The Internet is made up of a fantastic array of equipment and cabling that covers the Earth. Many groups, companies, and governments have banded together to oversee the protocols, standards, and network infrastructure that allow the World Wide Web to work. To a casual user the Internet is simply a connection point that allows them to access whatever tools and information they want to access. What happens behind the scenes of a wireless connection is beyond their comprehension, but the Internet is the largest and most sophisticated physical object humankind has ever built. It blankets the entire planet and is alive with information flowing to every corner of it in nanoseconds. As a tool to provide information the Internet is efficient, available at low or no cost to the user and almost magical in its ability to deliver with speed.

- Database systems, cloud computing, and storage platforms provide the foundation for storing huge amounts of information that can be indexed for fast retrieval. This might not be impressive if one hasn't ever architected a database; if you have, it is easy to appreciate the work and genius that has gone into inventing the hardware and software that allows us to archive immense amounts of information and make it accessible with a simple search request.

• Software applications provide specific functionality for delivering online learning in many different forms. We have brain training applications, learning management systems, and topic-specific online learning programs for free or for a small fee. Software provides simple ways for users to access raw information or structured curriculum in order to learn whatever they choose.

• Search applications continually improve their ability to know us and what we might want to see most when we request information. These intelligent systems are deceptively simple because either we just fill in an empty field with search terms or speak to an application verbally, then in nanoseconds our requests for information are fulfilled. Behind the scenes search functionalities are filtering our results in many ways. They are looking at us personally to guess at what we might be wanting based on everything the search engine company knows about us. This includes where we are physically. When we do a search for something like the cure for a fever blister, the system has to have the ability to sort through thousands of pieces of information on that topic in order to bring us what might be the most useful and accurate. And they have to do this task tens of thousands of times a minute.

• Mobile and wearable devices give us access to information at our fingertips at every moment of every day, while also pushing relevant information to us about anywhere we happen to be. The biggest advantage these devices afford us is the ability to ask a question and get an answer instantly. We do not have to wait to get home or work. We don't have to wait to go to a library or find a book somewhere else. All we have to do is type or verbally ask a question and we get an answer immediately.

• User-generated content Websites like YouTube, Slideshare.net, Scribd, etc. allow anyone or any institution to upload content and share it with the world. Each specializes in a specific format for information and serves as massive archives of information that can be tapped for entertainment or education. These are, in effect, the new-age library. Instead of books, they contain documents, videos, presentations, and in each case these alternate formats provide interesting and valuable information.

And the list goes on. Now we are moving well beyond face-to-face spoken words, books, chalkboards and even TV as the only ways to transfer knowledge. We are moving to a technology-augmented learning world where anyone who has Web access can choose to add knowledge at whatever rate they are willing to grow.

Historically we seem to go through a consistent set of phases with new tools from invention to adoption with the first phase involving many peoples' resistance of the new tools. After some time, the tools gain more general acceptance when it becomes obvious there is real value in using them. Momentum builds as people see others around them using the tools as a standard practice. Once they are adopted widely, often they become mandated, regulated and required so that we use them appropriately. In the education field we have followed this pattern a number of times since the days of the calculator.

When I was in junior high there were a few kids who brought a new device called a calculator to school. I had seen one because my dad was an engineer, so there was one in his briefcase though I had never touched one. Later that year we got a message from the school (on a piece of paper of course) that told us calculators were banned in class. By my sophomore year in high school calculators were allowed in school but only if you used the basic functions; using the more advanced features would be considered cheating. By my senior year they were openly allowed for use and even required in a few math classes. Our oldest child is named Kacie. When she was in school it was in an era where the desktop computer ruled. Therefore there was no problem with students bringing new technology to school. She was lucky enough to escape the pattern I experienced. She is seven years older than her brother, Austin, and when he was young we gave him a laptop; he took to it immediately. The Internet was young then but he knew how to use it to do email, find things on the Web and type documents. One day he wanted to take it to school and use it to take notes and look things up on the Internet. This seemed like a normal request to him because the school had announced

that they had installed the Internet; naturally Austin assumed he could use it. He was not allowed to use a laptop in class initially, though before he graduated laptops were commonly used in some classes. Our youngest daughter, Kristin, is a heavy iPhone user. She is seven years younger than Austin, and that separation is enough for her to have had the same kinds of experiences with her devices. The teachers make all the kids put them in a basket at school so they will not be used in class. Of course the problem now is that there is little difference between a tablet, phablet or phone, so where does the school draw the line on banning a mobile device? Wearable devices allow for the same options for sending answers to each other at testing time. Will we soon make kids strip themselves of all electronics before they sit down in class or will we accept that we need to teach WITH the tools and not by banning them?

With all of the technology tools becoming available to us for learning, we are improving and expanding the channels for gaining knowledge. We were limited at one time to learning on our own by reading books, talking to people or learning in a group through some kind of formalized class or course. When observing how learning is changing today, it helps to separate the methods into distinct channels.

Self-learning: Although we always have had the ability to check a book out of the library, read a newspaper or watch an informational TV show, now we have a much-expanded ability to use the Internet to learn about anything we choose. A person today could decide they want to be an expert in archeology, and there is enough structured Website content and self-paced coursework available to allow them to learn just about everything they would need to know. The best part is much of this is freely available to anyone who searches for it.

Group learning: Odds are most people reading this book grew up going to school and learning in a group environment unless they were homeschooled. What has changed is the ability to do collaborative learning online with people from all over the world. We can form our own *ad hoc* online social groups and share information at high speed. Often we work with teams of people from other cities in our jobs by using group email or online communities.

Matriculated learning: Most of us grew up with this form of learning as I've mentioned, wherein a person who played the role of an information "jug" poured out their wisdom into our "mug." This could have been a teacher, a parent or just someone who was older than us that we spent time with. Technology has widened the breadth and volume of people we can connect with in order to learn from them. It also has broken down geographic barriers because now we can communicate face-to-face with a person over video anywhere in the world. We have the option to connect with anyone we choose to pour into us. Sadly, young people often have little understanding of the potential of using technology in this way. They will connect with their friends but not with their elders. Yet the potential does exist, and hopefully as time goes on young people will come to seek out elders who can share their wisdom and who are valued for possessing it and their willingness to share it.

Flow learning: Because we have the ability to set up automated rivers of information with our technology today, we can have a filtered river of the exact subjects we would like to learn about pushed to our devices. By tapping into this flow, we have the ability to dramatically grow the volume of relevant information we can use to learn from each day. Twenty years ago flow learning came from radio or TV. Today we have online and mobile tools that provide the ability to pull or push information to us in real time. I have alerts on subjects I care about pushed to all of my devices so, regardless the situation, I can learn about the latest happenings on my laptop, phone, or smart watch.

Search learning: Because we have the ability to search for any piece of information in an instant with search engines and our devices, we never have to wonder about anything that interests us without satisfying that curiosity. Over many years, this ability will provide thousands of pieces of information to us that we would not have consumed in the past. This form of learning, which is very natural to younger people today, not only will provide many points of data to them; it will afford a confidence that they can learn about any subject they choose. With this confidence many will choose to become experts in subjects that would have been impossible to master in the past.

It is not just learning tools that seem to go through the same pattern of moving from resistance to requirement. I remember when pneumatic nail guns first hit the market. There were many construction framers who wanted to stay with hammers because they had been driving nails for decades and they could think of all kinds of reasons why the newfangled nail guns would be dangerous. Today you cannot compete as a framer if you use a hammer alone because the nail gun is 35% faster to use, and time is money. I wish we would just learn to adapt to new tools a bit faster and shorten the resistance phase.

5.14 million students are taking all of their courses in a physical classroom, while 3.55 million take all of their classes online.

Source: campustechnology.com

When it comes to learning there is a heavy price to pay when we delay the optimum use of technology. We stunt the possible growth of every person who could use these tools: kids and senior citizens alike. It's reasonable to wonder at this point how serious of an issue this is, actually. What if it just takes humanity twenty years instead of seven to figure out how to use technology at the optimum level of learning? That delta is such a short time in the context of the history of humanity. However, twenty years represents an entire generation that could have been more enlightened. This means potentially billions of people will not learn all they could have, and there are consequences to this. Often, ignorance is passed on in much the same way that enlightenment also can be passed on, and stunting the growth of one generation hurts more than we might think.

The faster we learn to learn (become more enlightened), the faster we gain wisdom. Wisdom can do many things for us aside from helping us make money. Wisdom helps us be physically healthier. Wisdom helps us better understand why we are here and what we are meant to be in our lives. This contributes to peace of mind. Wisdom allows us to be more intentional and on task and to give our intended gifts, inspiring us to flow love to others. Wisdom helps us be good models, parents and stewards for following generations. Wisdom allows us to use our tools and technologies in healthy ways instead of crippling ones.

Wisdom often often is associated with being older, and hence the phrase, "a wise old woman." If a person achieves an unusual level of wisdom at a younger age, we say she is wise beyond her years. My mother said to me many times as I was growing up that it really would help me if I would accept the wisdom of my elders instead of having to learn every lesson on my own, and in painful ways. Now what I realize about gaining wisdom is that it requires a combination of knowledge, experience and spiritual principles. Knowledge is made up of information we hold in our minds combined with confidence to apply that information. Experience is what gives us confidence in the validity of knowledge. To gain more experience we have to do things actively in life, and we do have some degree of control over the pace of doing things. To get knowledge we need a learning ecosystem that helps maximize what we can digest intellectually so we have the raw material to form knowledge.

Building a personal learning ecosystem is one of the most beneficial things we can do to help ourselves prosper financially and spiritually. Whether we are learning about the drivers of economics or about God, a learning ecosystem provides an engine of knowledge that will expand our minds and help us mature at a fast rate. A learning ecosystem can be free to build and to operate. If there was one thing I could get every young person in the world to do it would be to build their own engine of learning. In years past this could be accomplished by reading two or three books every week. In our time we have many more tools to bring to the party! In order to help you build a learning ecosystem, let me explain the pieces that need to be put in place:

1. Building a river of information: This is a carefully-chosen flow of digital information we consume in some way each day. We can sign up for blogs, follow experts on Twitter, read RSS feeds, search for subjects that interest us that day, and use alerts to drive new information on any subject to our devices. There are tools that will aggregate all of this as well as news feeds so that we have one or two screens to look at in order to see all the information available for us to consume that day.

2. Find educational social interactions: We learn from the people around us through the connections we have with them. These contacts can be social connections online or in person. Our friends share lots of information with us just in the discussions we have and the stories we tell. We make a conscious choice about the quality and volume of information we get from these sources.

3. Our work environment: Some of us spend a large part of our days working and our work environment can be rich with information flow, though not necessarily. In fact, most organizations are motivated to help employees learn because smarter people make better decisions for the company. We have some level of control over the quality and volume of this source of information.

4. Develop learning experiences: We can choose to experience many things that will teach us about life. We choose with whom we do things where we go geographically, what we do physically, and how many new things we will try. In may cases we also learn from experiences that are not of our choosing. Simply by being alive we will have opportunities to learn through the practical experiences of surviving. Life itself is a great teacher and some lessons can be quite painful. Painful experiences are not always a bad thing

because they activate us to reconnect with the Truth. Regardless of whether our learnings are wonderful or brutal, we gain knowledge at a high rate through experiential learning.

5. Apply reflective thinking: As we get all of this input we have to reflect on it and put it somewhere within our worldview. We have to give it meaning and context with respect to the other pieces of knowledge we already had. Without this step of reflective thinking it is hard to move from information to knowledge.

6. Update our worldview: There is Truth in the world; the wiser we get, the closer to the Truth we arrive. As our learning ecosystem hums along we learn more, either quickly or slowly, about the Truth of things. For this reason our worldview is changing all the time as we add more underlying knowledge to the computer that is our brain.

7. Repeat from step one: This is a cycle of course so we never stop learning our entire lives. Each day we stand in a river of information and do our best to digest its meaning so we can become wiser. At least most of us do.

Some people do not think consciously about the quality of their learning ecosystem; they just survive or live each day in search of whatever makes them happy. Of course they are learning a few things along the way, but not at a pace that really helps them jump ahead up the wisdom curve. Others have a passion for learning and will work at pieces of the process listed above. Very few are diligent about having all of the pieces in place in order to maximize their potential as learners. Online and mobile technologies provide powerful tools for helping us build and manage our learning ecosystems. Now we have powerful ways to assemble and manage our rivers of information so that we get better quality of information, and vastly larger volumes of it, into our minds each day. We have the ability to digest huge amounts of social information through all of the new social tools available. We have technology-augmented learning engines today if we choose to build and use them.

With all of these new tools for learning we must be realistic too that not everyone forms a learning ecosystem. Some people believe that the process of heavy learning stops when we get out of school. They move into survival mode as adults and learn at minimal rates, and primarily through day-to-day experiences. For these folks, all the technology in the world will not help them nearly enough to become more enlightened. In order to achieve wisdom, we must be fully committed to a lifelong learning ecosystem that does more than teach us through a few experiences each day.

Learning ecosystems get put to use in different ways throughout our lives. At the same time the technology involved also looks quite different. The learning tools that a 13-year-old uses are very different than what a 35-year-old uses. For that reason, let's consider learning as a lifelong journey and observe the different phases from young to old.

Young Child

Our new online tools and mobile devices probably are having a bigger impact on learning for this age group than any other. Whereas a young child in the past learned most of what they could from their parents or experiences, today they have tablets and mobile devices at their beck and call. Many parents today introduce their children to a small screen as soon as it can entertain the child. In many cases, this is somewhere between the 12 and 24 month range.

> *One in three babies has used a smartphone or tablet before age one:*
> "*The findings, which come from the responses of parents that attended a hospital-based pediatric clinic appointment between October and November 2014, show that smartphones and tablets are proving to be important pacifiers and distraction tools. 65% of respondents say that they employ the devices as a means of calming their children down.*
>
> *Nearly three quarters (73%) said that they let their kids play with smartphones or tablets while they get on with household chores and, perhaps of concern, 29% say they use them to help get their child to fall asleep.*"
>
> Source: nydailynews.com

The moment a young child can read, the connection they have to the Internet begins to play a role in their learning. Eventually there comes a day when they realize the device they have been playing with for a few years can answer their questions too; that is a huge deviation from days past. Before the Internet and mobile devices self-learning was much more difficult since a child only had access to their parents, others around them, and whatever books they could reference at home. Today they have all the tools they need to search and consume any piece of information they can think to search for. Many kids spend hours a day alone with their devices so they have at least the significant potential for self-education.

At a very young age new technology makes the process of learning a very different experience. These kids grow up knowing that virtually all knowledge can be gained for free and instantly at a touch of a button, and that is a very different place psychologically than we have been before as a species.

Adolescent

By the time they reach adolescence kids are in an expansive time of learning because they are schooled formally in some way. They spend hours each day having information poured into their heads as we attempt to fill them with knowledge that will help them succeed in life. At the same time we are teaching them how to learn and to value the process of adding to their knowledge, and this is just as important as what they learn. If teachers can instill a love of learning, curiosity, and the skills to acquire knowledge, they can set a student up to succeed for the rest of their life. These years are the critical first years of the formal education process and some students easily glide into this program while others do not. I was a "do not," by the way.

Schools have, up until just recently, used a memorization/regurgitation model at this age. In some cases, they still do use this method. Teachers pour information into students by presenting it to them verbally and through textbooks. If the student can recall the information on a test, they get a good grade. If they cannot pay attention in class and hate to do homework, they will struggle. This has much more to do with how well kids can memorize than how they can turn information into knowledge, however. This method of teaching is what created the paradigm of being "book smart" vs. "street smart." Street smart was most often the designation for a child who had a good ability to process information and learn through experience, and who did not adapt well to the memorization method delivered in schools.

Outside of the classroom our adolescents develop a deep understanding of the technology to which they now have access. It is used to help with their schoolwork of course, and they go well beyond that today. Their ability to mine information from the Web explodes as they begin to understand the reach and depth of information they can tap into at any moment. Their friends inundate them with links to Websites, new mobile apps, links to content, and online content providers who might be interesting to them. Because they cannot drive themselves anywhere they have hours at home to discover information from their various screens.

Although information is freely available to them, Young people can also become complacent about accepting the first answer provided by a search engine without having the skill or understanding of the importance of checking the source or finding verification and proof of the information presented to them.

Online information is being used today by adolescents to supplement what they learn in school at an extremely high rate. They spend less time asking their parents for advice because, frankly, they can get any piece of information instantly and it will be more expansive than what a parent provides, most likely. Some of the learning that adolescents want to do is on subjects that are embarrassing for them to ask about (sex, drugs, alcohol, and any vice they run across), so much of this learning is done through research online.

The important new consideration about learning that we need to understand for both of these young age groups is that the learning boundaries are gone now. They can learn about whatever they want and as fast as they can consume information. Which subjects they choose to self-teach will dictate heavily what they become as they grow up.

Teenager/Young Adult

By the time most young people reach the age of thirteen today, they are deeply familiar with the Internet and the devices to access it. They have become quite addicted to their mobile device because it provides a connection to their friends and provide information on a majority of the problems they face in life. It entertains them and provides answers to any question they have. Their addiction to their mobile devices is understandable considering it is no longer just a phone; it is a tool that represents a huge part of how they operate in life. Though they can get along without it, they feel crippled when it is not available, and in some ways they really are crippled. With that said, I still talk to some teenagers who enjoy being away from their mobile devices from time to time, though just for a few hours from time to time.

92% of teens report going online daily — including 24% who say they go online "almost constantly.

Source: pewinternet.org

Schools are moving now to new learning models like the flipped classroom. This is best described as a teacher sending students home with videos to watch online to learn a subject and then doing the homework in class. This works well because students can do the learning part at their own speed and can repeat the video multiple times if needed. In class once the content is delivered, it is done and there is no rewind. Then when the students do the homework in class, the teacher can help them make sure they have understanding. This is as opposed to doing the work at home where there may be no one around to help if the student has a question. So is the flipped classroom working?

- *In 2012, 48% of teachers flipped at least one lesson; in 2014 it is up to 78%.*

- *96% of teachers who have flipped a lesson would recommend that method to others.*
- *46% of teachers researched have been teaching for more than 16 years but are moving towards flipped classrooms.*

- *9 out of 10 teachers noticed a positive change in student engagement since flipping their classroom (up 80% from 2012).*

- *71% of teachers indicated that grades of their students have improved since implementing a flipped classroom strategy.*

- *Of the teachers who do not flip their classroom lessons, 89% said that they would be interested in learning more about the pedagogy.*

One important distinction to make about flipped classrooms is that not every subject needs to take on this approach. It is better to start with just one or two lessons. The traditional approach still has merit and certainly should be utilized. Consider the flipped approach as a creative way to supplement learning and foster student engagement with the content.

Source: Sophia.org

This is a great example of how humanity and technology are blending into a Humalogical balance. We are moving into a wonderful time of blending when a teacher can blend with technology tools in order to improve how learning is delivered. Instead of the learning process being done primarily through a live person sharing information verbally and through books, now we have technology to deliver content, with the teacher acting more as a guide in the learning process. Learning still can be done through a live teaching model, and now with online tools as well. Both have plusses and minuses. It is likely we will spend the next decade or so experimenting in order to get the most effective blend for an age group as well as discovering how best to accommodate various types of learning styles. No one learning model works for everyone which is important to understand. There are people who learn better from another human being and who cannot digest online learning quite as well.

46% of students' skip classes when course lectures are available online.

Source: ecampusnews.com

This age group also is under much more pressure to be voracious readers than their younger compatriots. As a teenager the coursework moves toward a much heavier load of reading on one's own in order to learn; this just keeps increasing in college. Technology is changing how young people are learning to read, and not in a positive way. Because these young generations spend so much time in front of screens they read more by volume than people in the past, though they tend to read in small bits and bites of sentences rather than in long page-after-page blocks. For example, a teenager might read 10,000 words a day with much of it coming from reviewing the posts of friends online. A teenager in years past might have read 5,000 words a day, most likely from a book, magazine, or an occasional letter. Because young people today read in small blocks, they tend to struggle with the concept of reading a long book. They can read an interesting novel that is entertaining but when they are faced with reading full textbook material, many really struggle.

When my son Austin was in high school, I walked into his room one night while he was doing homework on his computer. He was working on a twenty-page term paper and, upon hearing that, I was transported immediately back to a day when I was in high school and had to do a twenty-page term paper on Mahatma Gandhi. I distinctly remember dreading the fact that I had to write out twenty pages by hand, which seemed to be a task beyond my level of

149

patience. For days I sat in the library with the three books they had on Gandhi and read, then re-worded what the books said about him in order to tell his story in something akin to my own words. After what seemed like weeks, which actually was three days, I had the opus done. I got a C. I asked my son to show me how he was putting together his report and he showed me how he went online, did a search on the topic and began to cut and paste lots of text from different sites and online documents into his own Word doc. Then he went back, reorganized things and changed the wording about 20%. I asked him why he reworked the content 20% and he informed me that the teacher would be running the term papers through an online engine that checks for plagiarism. A few unpleasant thoughts ran through my mind about the tools he was getting to use, and his ability to access information as opposed to mine when I was his age. Suffice it to say I was more than a little jealous. He got his report done in part of one evening, reworded things just like I did and, for the same reason, and he was watching TV on his computer the whole time. The biggest blow was that he got an A on his report.

Because learning for this age group is so heavily influenced by school, let me make a few observations about formal education. Now we are in a painful process of transition where we must reimagine what the purpose of school is in the technology era. When students have such a far-reaching ability to learn anything they want online, and since there is better content available there than any one teacher could deliver, we have to adjust our thinking. The reason this is painful for some administrators and teachers is because we have been operating within a model that was developed at the turn of the last century and many of the habits we fell into were more to help schools operate fluidly, not necessarily to help any one student. I am sure this is the reason we are seeing so many alternative model schools springing up.

In order for schools to be relevant in this era of technology-augmented learning, they will have to focus on being strong as institutions that provide:

- *Social maturing and a focus on EQ as well as IQ*

- *Technology-based self-learning skills*

- *The skill of building self-assembled Rivers of Information*

- *One-to-one learning that can customize learning at a higher degree*

- *Facilitating student-discovered learning instead of direct information transfer, allowing students to discover information instead of handing it to them*

- *Less fact-based learning and more critical thinking, creative problem solving, discretionary thinking and experiential learning*

Of more than 6,000 students polled across 36 campuses, 77% said their grades improved through web-based course material and online classroom managing sites like Moodle.

Source: mndaily.com

Adults

We live in a world where the pace of change is speeding up, and with this comes a greater necessity for learning. When change is really slow we are under less pressure to learn about new things, and the knowledge we gain will stay valuable for a longer time. However in times of great change, it is important to be adding knowledge constantly in order to be able to provide high value in the new environment. This is putting pressure on adults today to have an effective learning ecosystem. To have that we must rely on technology to provide a flow of information that is timely.

There is a clear dividing line when we got the Internet and mobile tools, so there are some adults who have a lot of experience today with using them to self-teach and older adults who grew up with books as the primary medium for conveying information. I am an adult that is split down the middle in that I spent half of my adult years using paper books as primary learning tools, and then the other half using the Web to do searches. Books are wonderful for delivering in-depth information on a subject. They have one drawback, however; normally the information is dated by the time someone would read it. One of the great benefits of getting information from online sources is that we can tap into the very latest information on any subject. Most adults today are pretty comfortable with consuming information either from books or the

Internet because we grew up with books and have had plenty of time to get used to the Web as a resource.

As always there are positives and negatives to getting information from online sources. A book has to go through a publishing process with an actual publisher or through self-publishing, so there is a pretty big commitment to the content being valid. A Webpage can be published easily and it takes a much smaller investment of time. So there are thousands of people who might "publish" information on the Web without much of a filter, so the efficacy can suffer. Second, search engines often filter the responses for our searches. They do this to "help" us by narrowing the results we get to results they think we might be interested in. The problem, of course, is that they are narrowing our field of content artificially to learn from. If a search engine knows where we are geographically, we will get information that is close to us physically. But what if we wanted to get information on our inquiry from around the world? The danger here is that a search engine can influence what we are exposed to and, thereby, what the base of knowledge is that from which we learn.

I learned a very good lesson about the filtering of information as an adult when I went to the Soviet Union and Germany and saw how they viewed the Second World War compared to how I had learned about it in my filtered U.S.-based education. In Moscow and Leningrad (what it was called when I was there) I saw many war memorials, so I asked my hosts about them. I learned that more Russian people died than any other nationality in the war. I learned that almost a third of all of Leningrad died of starvation during the German blockade. In the U.S., we are taught about how many Jewish people were killed in concentration camps, but not how many of the "enemy" forces died, or even some of our alliance partners. While in Germany, I was discussing the War with a German businessman and asked him how the nation has resolved the legacy of Hitler, and what he and some other German leaders commanded during the war. He looked at me and said, "How is what they did any worse than what the Americans did to your Indians years before? You took their lands from them, systematically killed them off and, to this day, you have marginalized them, though they were the original inhabitants of your country." At that moment I realized that I

was never taught how many Native Americans we killed, how we killed them off, why our government did this on purpose, and that what we did was akin to genocide. His point was well taken and hard to admit as an American. Victors write history and I have never forgotten the lesson I learned about the dangers of sanitizing or filtering the Truth. Now I realize that my education as a young person was delivered through a filter that was chosen for me and that I accepted as the truth without question.

Because search engines also are filtering what we see when we search, and for many of the same reasons as in my story, online filtering can be dangerous. Someone else is in control of the information we are exposed to.

For an adult to be successful in the economy today there is a constant process of learning and re-learning that must go on. Obviously new technology allows us to learn whatever we choose, and virtually for free. We have the ability to be lifelong learners, and so we have no excuse at this point. The cost is pretty low for devices to connect us the Internet. The available volume of information on any subject is tremendous and growing daily. The only things needed are the curiosity and will to learn, and any adult can set up a sophisticated River of Information with free applications that make content easily available. This is a wonderful new moment for humanity because it levels the playing field economically, socially and practically. Anyone anywhere in the world who has a desire to learn about a subject is no more than a moment away from starting.

From this point forward more people will discover that they have complete control over what they learn about and how deeply they become an expert about a subject. As they make this discovery they will use their new knowledge to prosper and grow, while others will see this and be motivated to do the same. I predict this will cause an impressive worldwide escalation in knowledge over the next three generations that will yield solutions to many problems that would not have been solved in the past. It will free the underprivileged to have access to the one thing that can make a huge difference for them: complete control over their knowledge.

Even though technology-based learning may be a blessing, there are dangers lurking like the possibility that we will depend on our online systems to do our thinking for us. When information is so easily available, and when there are so many people creating content that encapsulates their knowledge, lazy thinkers might succumb to "group think" instead of forming their own opinions by sorting through the underlying information. In some ways this is what can happen with elections today. Many people do not take the time to understand in greater depth what a candidate stands for, or they may not even listen to a debate. So they make a decision on their vote by listening to a few talking heads in the media.

This can spin out of control as we have better and faster access to people posing as experts; we could slip into doing what they tell us to do in any given situation without understanding the underlying reasons. As the Web gives us better tools to enhance crowd dynamics, it will get more enticing to ask the crowd what they think and go in that direction. Or we may look at a rating site and do whatever the raters say is best. We can get in trouble when we skip the step of learning underlying information on our own, trusting another's knowledge blindly instead without verification. This will be exacerbated as we get new technology that is right around the corner; they will give us easier access to the opinions of other people and organizations.

New devices and capabilities will continue to give us wonderful options for gaining information to enrich our learning potential. Wearable devices will move the Internet closer to us physically. Brain-computer interfaces will let us control devices just by thinking. This will allow us to interact with technology at the speed of thought and, consequently, to speed up the process of learning about anything we choose. Decision support systems will offload our responsibilities to gather and review lots of information in order to make decisions. Of course this means that, in many cases, we will not even be able to make decisions on our own without these smart systems helping us. Holographic teaching systems will allow us to blend online learning with stand-up delivery so that we can learn in ways that are best for each of us. They will allow us to learn about any subject from a teacher who is the top expert, and not just one that is near us physically. This is just a taste of the kind of technology that already is emerging in our world, and this does not even include the inventions soon to come that will deliver information to us so we can learn even faster.

We have crossed a line where learning is much more under our control and always will be from this point forward. We do not have to depend on a teacher, parent, school or university in order to learn about anything we choose. We do not have to move to another country or even borrow our future away to learn what we need to know. Now we have the tools and content to learn about whatever we want as fast as we can consume the information. This is a giant leap forward for humanity if we will take advantage of it and invest our minds and time in moving data and information to wisdom.

SUMMARY:
Digital Learning

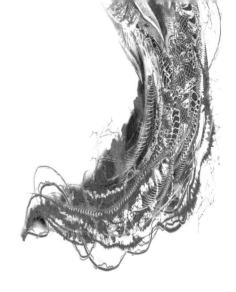

I could make the argument that there are not many things more important in our lives than the ability to learn. From the moment of birth until the day we die we have the opportunity to learn and, in my opinion the responsibility, to learn and grow. The amount of knowledge we can gain is infinite because there is more information created every day; it never ends. If we are careful about what we learn, we can marinate our minds in information that will enlighten us about the Truth of things. This applies whether we are talking about career matters, life situations, or spiritual discoveries. To the extent that we are willing to search for new understandings, we have an unprecedented set of tools now to seek the Truth. Word-of-mouth and books in a library are crude tools compared to the Internet. The Web provides an expansive conglomeration of information that humans have archived over our existence. This, along with our new mobile devices, provides unfettered access to an unimaginable volume of knowledge. The only tasks left to improve humanity are our collective and individual wills to learn on a constant basis.

My light versus dark ratio for digital learning as a tool for humanity is 70/30 toward the good. There are dangers with young people learning about parts of life before they might be ready. And there also is the untrustworthy nature of some content online that can skew what people learn and then believe to be the truth. However, this pales in comparison to the positive nature of being able to learn vast amounts at low or no cost, just about anywhere in the world. Knowledge is a wonderful thing and digital tools are growing humanity's knowledge base. Wisdom is even more powerful because as we become more enlightened we move to a place of maturing our world toward a place where love, joy, and peace are the norm. Yes, this sounds utopian, but I believe fully it is possible.

Humalogy Viewpoint

From a Humalogy viewpoint the education process has shifted slowly from what I would estimate was an H4 to a score of T1.

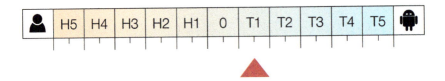

I get to this by observing how much the people around me in schools and universities spend time with stand-up professor led delivery versus eLearning. I then have to factor in people watching videos on YouTube, which rates a zero on the Humalogy scale (equal parts human and technology). I do not believe that education will ever become a T4 or T5 because that would mean that a computer is creating the content and deciding what we should learn, and that has dangerous implications.

Having a human involved with the learning process is helpful because we have a high level of discretion and empathy for each other in the learning process. Although technology-augmented learning is a fantastic driver of growth for humanity, learning does not mean we will be more enlightened. We may have more knowledge, but whether we translate that into wisdom is a choice we have to make. Surely if God loves technology this would be one of the reasons we can translate our newfound knowledge into enlightenment and use it for positive and healthy uses. We all have a choice about what knowledge we accept into our minds; that could be how to build a bomb or how to ease another's depression.

Being an optimist, I am hopeful that our technology-augmented ability to learn will be used for learning things that move the world forward and not drag us backwards. One thing we know for sure: God values wisdom because we have many verses in the Bible that exhort us to be wise, so it stands to reason that our new tools are positive. It also is pretty clear that those who believe we should go back to traditional learning views such as not educating females (still a belief in some cultures) clearly are standing in darkness.

Seminal Questions

Will the majority of humanity seek to build powerful learning ecosystems and thereby accelerate our quest of knowledge and wisdom?

How should we handle the fact that young people can self-learn on any subject, including those that they may not be mature enough to understand?

In what ways will our maturation levels as human beings increase with our new abilities to learn through the use of technology tools?

How will the formal education process continue to change as we have the tools to self-teach at higher abilities?

How will the ability to learn with free online tools change worldwide economies and standards of living?

Will we harness technology-augmented learning to become more enlightened, or simply defer to our machines as knowledge sources because it is easier to allow them to think for us than to learn for ourselves and make our own decisions?

CHAPTER SIX:
Frictionless
Communication

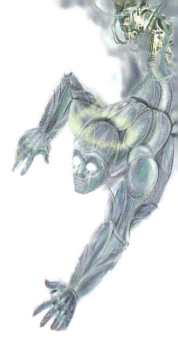

"Changing how people communicate will always change the world"

—Quote from Facebook Corporate

Something as simple as having a conversation with a person plays such an important role in our lives. We are already forgetting that not so long ago it was very difficult to talk to some of the people we wished to connect with, either because they were on the other side of the world and it was too expensive or because we simply did not have their current contact information. Over the past ten years we have dramatically improved our abilities to communicate one-to-one, and one-to-many. The costs have dropped almost to nothing and the convenience of finding people and communicating with them has much improved. Sometimes we fail to recognize how changes to basic things in our lives causes monumental shifts.

The ability to communicate without friction is changing many important dynamics in our lives. We are getting smarter because information is shared at a much faster rate between people either one-to-one or one-to-many. We are becoming less concerned about privacy because in order to communicate with people around us, we are allowing them to know more about our status and our lives. We are less concerned about distance being a barrier to communication with others. Whereas it once was expensive and unwieldy to talk to someone across the planet, today the most difficult challenge is managing the time zones. Most important, because we can communicate so easily at this point, anyone with an Internet connection can be an information provider for all the rest of us. Rarely do we stop and think about how our new ways of communicating change us; we just get excited about the next new tool that makes it easier or more convenient to communicate.

In this chapter we will focus on the specific channels and tools through which communication can be shared. In the formats of text, video, graphics and audio files, we digitize pictures, our words and thoughts so they can be delivered across a room or over great distances. We digitize them so we can shift time and "talk" to someone in something other than real time. We digitize our ideas and information because we can deliver them inexpensively and so we can talk to tens, hundreds, thousands, or millions of people instantly and with a keystroke. And lest we forget: we have only had this capability in general use for the past couple of decades.

"54% of texters state they text their friends once a day, but only 33% said they talk to their friends face-to-face on a daily basis."

Source: nytimes.com

Communication as a system can be looked at as people flowing information, emotions, ideas, art, news, needs, answers, and other content to each other. This flow has grown from a tiny trickle when we did not even have a spoken language to communicate. Now it is a wide-open spigot, replete with streams of communication that happen now each and every second. The volume of communication between human beings is growing exponentially at this point. A frictionless, free flow of communication is going to be yet another huge impact of the Internet and our digital tools. As always it is important that we learn how to leverage what is positive and minimize what can be destructive.

From a spiritual viewpoint transmitting ideas and information to each other through communication always has been the key to allowing those who are wise to share that wisdom with others. This is done through conversations, preaching, counseling, and committing wisdom to writing. Without communication it is very difficult to spread wisdom because it must be learned anew with each person. With technology-augmented communication it will be much easier to share knowledge and wisdom. God communicates with people in many ways: through words, in dreams, with voices in our minds, and through other people. Is it possible that the Internet will turn out to be one of the most powerful tools for spreading knowledge and wisdom by enhancing our ability to communicate with each other? Deep in our soul we are called to holiness. Most of us struggle to align that call with the cultural world we live in. It is my hope that technology will play a role in helping us more easily align what our soul calls us to do with the earthly life we live. In order to do that it will need to enhance the ability of wise people to share that wisdom with the rest of us. Already there are signs that this is happening.

We are reducing the amount of friction involved in communicating rapidly. Just over the course of my lifetime I have experienced a dramatic change in how I relate and connect with those around me, and with the world in general. In order to have a historical perspective, let's dive a little deeper into the stages and timing of communication advancement from centuries ago, then work our way forward.

Approximately 2,000 years ago, the primary form of communication was word-of-mouth. The relatively few documents that were written in a physical medium were written by hand by one of the few people who could read and write, and were stored on some form of papyrus, clay tablets, or other flat surfaces that rarely lasted very long from a historical perspective. Mass communication normally consisted of pictures or a few words carved into stone on buildings or monuments, or delivered by a person at the top of a hill, in a square, or on a pedestal. Communicating over distance was a problem because the physical media for words did not travel well, and if people carried the message it was difficult to keep the communication complete and error-free while moving from place to place. Over the course of 1,000 years not much progress was made, although we got a little more proficient with making paper and binding books that were written or copied by hand.

Leap forward to the printing press: we invented a powerful solution for solving the distance and archiving problems because we gained an ability to mass-produce words on paper. By having the ability to reproduce volumes of words with a machine fairly quickly, we could inexpensively and quickly reproduce books and fliers. This brought the availability of books to most everyone who sought them over time, and made it more likely that the information in the book could be communicated over a great distance and to many people who would not have had the chance to learn through reading.

The impact 600 years ago of printing press technology can best be understood through the changes in society when the Bible was mass-produced. Instead of the word of God being delivered only by priests it was possible for a common person to own a Bible and read if for themselves. Although today we take for granted that we can have a Bible in our hands in book form or on our mobile device, before the mid 1400s, people had to rely on word-of-mouth to receive biblical wisdom. The danger with that is that any human who interprets the Bible could (and did) make mistakes as to what the Bible might be saying on a specific topic and be misled. So in this case note that the Guttenberg Bible was the first book to be mass-produced and that was a critical early use of what was a new technology at that time. Maybe this is symbolic of the role our modern technology will play in expanding our spiritual awareness and understanding.

So radical was the concept that a common person could own a Bible and interpret it for themselves that some church leaders fought this movement made possible by Guttenberg. Egotistically they knew that to lose their status as the sole source of interpretation of the Bible meant losing much of their power. Entire denominations grew out of the capability of a person to own and read their own Bible. Later the printing press would spread thoughts and ideas that changed nations, caused the downfall of kings and governments and drove the desire for millions of people to learn to read so they could understand the words and ideas on the pages of many printed books. Compared to advances today the printing press is a crude technology, but inventions need not be earth-shattering in complexity to make a huge difference in communication.

For hundreds of years after Guttenberg, there was not a big step forward with communication capabilities. Not until around 1837 did we learn to harness an electronically-aided form of communication that offered two valuable new means of communication. The telegraph and Morse code gave us the ability to communicate over long distances in real time and allowed us to send messages person-to-person for the first time through means other than a paper and pen.

This was a huge step because it ushered in the first time we would use the combination of an electromagnetic transmitter and a form of digital communication (Morse code) to improve on written words. As with any other big step forward it seemed magical at the time. People could hardly understand how, by using a simple wire, two people separated by many miles could actually talk to each other. The telegraph was not an efficient way to communicate because people only could afford to send a paragraph of words, and they did not know for sure that messages reached the person at the other end. If the message did reach the recipient the sender would not know that for sure unless someone responded to verify receipt. If the wire was cut anywhere in between stations the system was inoperable because there there was no redundancy. Finally, there had to be people on both ends who knew Morse code or there was no way to decipher the taps being sent.

An interesting example of history repeating itself is that, with Morse code and the telegraph, a person might only send 140 characters in order to be cost-effective. This caused of abbreviated language that had to deciphered in order to understand the message since all extraneous words were left out. Now here we are, centuries later, doing the same thing with Twitter though this time it is by design rather than necessity.

Move forward just a few decades from the invention of the telegraph to the 1870s and we have the roll-out of the telephone. Though it was invented at least twenty years earlier, but the telephone was not incorporated into general use until Alexander Graham Bell brought it to market with lines and switchboards. By 1880 there were 49,000 phones in use and the world was about to take another huge step forward with communications. This time a person could talk directly to another person in real time and with actual words, tone and emotion. This time there were people that would hear the voices coming through the "technology" and believe not only that it was magic; some even believed it was the work of the devil. Once again proving a dynamic I will mention a few times in this book and that we have seen with humanity in general: we often fear new inventions and systems until they are normalized and accepted as part of culture.

As we all know the telephone exploded in popularity and became a must-have piece of technology for every household. Over the next 100 years phone networks grew to include every corner of the world. Phones themselves improved as well. The equipment that handled and routed the calls improved annually over that decade. Although it was relatively expensive to talk to someone across the world in real time it did become possible. This sped up the flow of news so that any happening anywhere could be reported on within hours of the event. This allowed family members to stay in touch, even when they lived thousands of miles apart. The telephone shrank the world in many positive ways. The only two big pieces of friction in this system were the cost and the fact that we had to be hard-wired into a phone system.

This would change with the advent of mobile and cell phones in the 1980s. Once again we saw another element of friction overcome because we did not have to be at a building (home or office) to receive or make a phone call. Life changed again because we could communicate with someone regardless of where we or they were at any given moment. This brought many changes to society because carrying a device that allows for instant communication means people are available to their family, friends and coworkers whenever one of us feels like reaching out. If a mother wants to stay in touch with her child, the situation is mostly positive and healthy. If it is a company contacting us as workers all night and over the weekend, there can be negative consequences.

The other giant step forward birthed around this time was the Internet. The concept for the Internet was to create a separate network that was not focused on voice, but was focused more on moving text from point to point. The Internet was first through of and organized in the 60s as a method for connecting large mainframes from different organizations so they could share information. The Web as we have come to know it came along in the 90s when we extended past simply having text on a screen and added the capability to add graphics, sound and video. At first the Internet shared the telephone system backbone in order to allow for the communication with text, sound, and video, but this meant using a modem to "dial up" onto the Web over a phone line. Later, companies built separate digital "plumbing" in order to deliver the Internet.

Once the masses had access to the Web, the number of different communication tools began to explode and are still being invented to this day. We got email, chat rooms, instant messaging, text messaging, online communities, Skype, Snapchat, Whatsapp, and the list goes on. In each case the application filled a specific role in the communication toolbox. And as you might have noticed, each of these became indispensable to various groups of people. Try to convince a teenager to give up texting; you're likely to be about as successful as if you tried to get a company to shut down email.

What we have done by digitizing our words is create a near-frictionless way of delivering our communication to one or to many. Compare, for example, the friction involved in sending a letter or document through the mail. We had to memorialize our words on paper, which takes a huge amount of resources to create, ship, store, or destroy. Then we had to "print" on that paper in some way with a pen or a printing press. The letter/document then needed to be folded or put in an envelope, and then had to have postage purchased and applied. Only then do we start the resource-intense process of delivering the document/letter to a specific box somewhere else in the world. And the real problem with this process is not just the friction to get words delivered; it is the time that it takes to get the words delivered, which can take days instead of moments. This is a high-friction system that we have been using as one of the options for quite a while, and it is still in use although shrinking at a steady rate.

Communication has shifted on the Humalogy scale toward the technology side. Whereas long ago we communicated completely through storytelling and face-to-face communication, now we have shifted way over on the scale to a T2 for most people. This means that a majority of the "conversations" we have today are through email, text messaging, online posts, and other web-based and digital media. The reason for this is not that we all love technology; it is because the digital tools have taken so much friction out of our traditional means of communicating (face-to-face, telephone and letters) that they are indispensable for today's way of life.

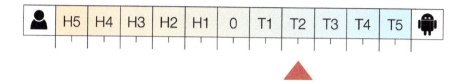

We have invested a lot of time and resources into getting to the point of having frictionless communication, and we are not done yet. If cave drawings and smoke signals are at one end of the spectrum, direct mind-to-mind communication would be at the other end. The ultimate form of sharing information would be to connect telepathically anywhere in the world. This can be a scary thought because we all have too much going on in our minds to process on some days. Giving people the ability to talk directly telepathically is a little hard to imagine, so in order to stay on track with what technology can do let's just assume that at the far end of the progressive side of the spectrum we will have the ability to have technology that allows us communicate instantly.

Let me dispel some thinking that seems to be going around about new forms of communication killing off all that came before. Just because we are widening the pool of communication options does not mean we should not assume our "old" ways of communication will go away. There is room for many different channels of communicating because most of them have unique attributes that we value. It is naïve to think, for example, that text messaging ever will replace email. Email gives us an ability to "talk" to someone with a relatively short question or message, though not in real time. Texting most often is used for a real-time conversation and is limited to a small number of characters.

Now we have invented new ways to communicate that are virtually frictionless. We can talk to millions of people in real time with the click of a button and the cost is minimal now. With each leap forward, just as with the printing press, new communication tools bring changes in behavior and collateral impacts that change us as human beings. We become better informed. We build relationships differently and with different people. We don't need to travel to the ends of the earth in order to communicate in a robust way with people anywhere in the world.

The evolution of online communication capabilities has blended the Web as a transport protocol with wireless capabilities in order to give us freedom from space and time. We are unchained from any barrier to communicate whenever or however we choose. Being unchained is changing some interesting things in the world:

Relationships grow and change when we have a vast array of communication channels: With a frictionless ability to communicate we find new people to communicate with. It is much easier to communicate in general so our abilities to transfer information to others is greatly enhanced. We have tools to send short bursts of information, pictures, or video, all of which enhance the visual dialogue of our communication with others. Quite literally we can document in real time what we are doing or thinking and we can direct that to a group, to a specific person, or to the world. This all adds up to an ability to manage more relationships in a smaller amount of time.

The ability to communicate within a relationship can be immersive if allowed by both parties. This has the potential to speed up the depth of the relationship and to allow a relationship to grow regardless of distance.

My wife Annette and I began to date after email was available to us but before texting, Skype, or Facebook were in the world. Just with email available to us to augment the telephone, the way our relationship grew was impacted greatly. We worked at the same organization but lived 100 miles or so apart. Generally during the first three or four months of getting to know each other, we used email to share our past, our feelings, our hopes, and our dreams. Sitting for hours on the telephone was not possible but sending email back and forth when it was convenient to write was great. I am not saying we would not be together without having email as a communication tool, but it did play a major role. Jump forward to our children today, it is clear that a vast majority of the communication taking place between two teenagers who are dating is done through social tools and text messaging. In fact I would guess that young people who are dating and go to the same school still spend more time communicating electronically than they do in person. It is very normal today for people dating and living some distance from each other to use online tools as the main source of communication.

Communicating with online tools gives people digital courage: It is much easier to write something to someone from a distance when we do not have to face the emotions that what we say might generate. When we are face-to-face, we risk experiencing the fear, embarrassment, or rage of the person with whom we are communicating more intensely, immediately, and perhaps unpredictably. We can say things to the world through an online channel that we would never be able to say out loud in front of a group. Commenting electronically isolates us from some of the consequences of a response we do not want to deal with.

There are benefits and drawbacks of digital courage. One the downside, digital courage leads people to say things that can be overly harsh and nasty without the fear of accountability. Because we do not have to face the reaction in person we have little to hold us back from letting loose. Of course the fallacy with this is that whatever a person says electronically is stored forever in some way and can come back to haunt us later.

On the positive side the ability to talk to someone electronically can free a shy person to express themselves in a way they rarely could do in person. Husbands and wives can share feelings or facts that might be embarrassing in person but need to be shared.

When I was young I read the Diary of Anne Frank in school. This true story written by a young girl hiding from the Nazis has touched people for decades. The problem is that she died before the end of the war and her story only got out afterward. In today's electronic world she could have been a blogger or someone would have helped her get out of hiding. Perhaps her story might even have had a better ending. She had the courage to write her story but did not have the channel for people to read it instantly across the globe. Today she would be able to talk to all of us through many channels and we might even have responded in time to save her.

Today people have an easier time maintaining relationships when there is distance between them: Before email it was expensive to stay in touch with a person by phone who was distant geographically and and it is challenging at best to stay in touch by mail. I started a company with the Soviet government in 1988. Once a month or so we would try to arrange a phone call so I could talk to my Russian partners. Not only was it expensive; it also was maddening to even get a phone line between us at a predictable time. Snail mail always was an option but it was impossible to run our business through the postal service.

The example of this that tugs at my heart more than any other is the ability of soldiers to videoconference with their families from any part of the world. Every generation of soldier before the present era was forced to communicate with letters and a rare phone call. This meant when soldiers went overseas on a mission, they were lost to their families for months at a time in most cases. Putting their lives in danger for our country is just about one the noblest things someone can do. Giving up daily communication with one's spouse, kids, and extended family members makes it even harder. Today soldiers have the ability to stay in contact from 10,000 miles away and that capability should not be taken lightly.

Today we have everything from text messaging to FaceTime and Skype to help us keep in touch all through the day. We have a choice to communicate through text, videoconferencing or even through still photos or videos. Distance does not matter in the cost equation any longer since the cost is bundled into our data services, or is "free" to us if we are using others people's bandwidth. As free bandwidth becomes more pervasive - and it will - it will get easier to communicate with anyone, anywhere, for free.

Our daily awareness of the activities and lives of family, friends, and strangers is growing: As more people around us communicate through online tools they build an inventory of artifacts about their lives. These snippets of information paint a more vivid picture of daily life than we have had in the past.

There are people online who over-share and there are those who never will share. A lot of people are somewhere in between. The trend over the past ten years certainly is showing more people using online and social tools to communicate with individuals in their lives or with their online friend groups. This has created a tsunami-like flow of online information being posted, sent, and digested that helps people communicate one-to-one, or one-to-many. We know more data about more people than ever before and this impacts our behaviors in many ways. When we know someone is hurting because they posted this information online, people take action to help. When a person's special someone changes their relationship status online the other party gets the point that a break-up just happened. When we see a touching video about something that happened to a stranger we are impacted.

The impact on the human race of this level of communication already is enormous and it will continue to grow. As it becomes more common to automate the process of updating people around us the volume and frequency of updates will grow. When we use wearable devices we are able to automate informing people about where we are and how we are doing physically. With heads-up displays we are able to record anything we are looking at on the fly and send it to any person or group we choose. How we define privacy, or at least what we consider private, will continue to evolve (we will talk about this in detail in a later chapter). It will seem decreasingly strange to automate a portion of our communication so that those around us have a deep sense of what we are doing. Recently we added to our inventory of communication tools the ability to stream real-time video through Twitter so we can "report" on an activity for others to watch.

When information flows through communication channels more freely our knowledge levels improve: The more communication channels we have to use and consume, the more volume of information that can flow to us. If a person does nothing other than spend an hour a day reviewing posts on their social sites, they will gain large amounts of information over time. There are many people who get the majority of their news from reading social sites.

> *"Three out of ten American adults get a portion of their daily news when browsing Facebook. In addition, more than three-fourths of the respondents see news on Facebook when browsing their feed for other reasons and 22 percent believe that Facebook is a good place to read news each day. Interestingly, more than a third of respondents like a news organization or specific journalist on Facebook, thus social updates from those pages are laced into their news feed."*

Source: Pew Research Center's Journalism Project

Some of what we read in our social streams is mundane and useless, and some of it is valuable. Think of all our new channels of communication to be like new windows into our minds. We once had word-of-mouth and paper-based communications as our only windows in, so the volume of information flowing to us was restricted. Today we still have those two, and have added a vast array of channels that flow through our mobile devices, social technologies, email, etc.

Our new communication channels not only improve the breadth of information we get; they also improve the speed and depth. In many cases I learn about some new piece of information because it is pushed to my phone where I consume it hours before I would have received it in the past. In our personal lives, being made aware of a friend in need at the moment they have a need gives us the ability to respond much faster than in the past.

Even more interesting is our ability to add depth to the information that comes through all of our new online channels. When we communicate digitally through tools like blogs, articles, posts, and news feeds, we have the potential to provide links that can lead a person into a deeper level of information with a quick click. Even without a link to prompt us, we are able to do an online search if a piece of information causes us to be curious, and we can learn as much as we care to by investing more time browsing around the Web. In the past our ability to "search" for information was restricted to asking people around us and, in many cases, they would not have had it. Now we can poll the world in real time in order to dive deeper into whatever intrigues us.

Many people tell me they feel overwhelmed by the amount of information delivered by our new channels of communication, and I get that. As with any new tool there will be a learning period when slowly we discover the best ways to engage with our communication channels in order to maximize what we want out of life. We will get better filtering tools over time too. It is a natural first step for our new technology to overwhelm us with new capabilities. The second step will be for us to adjust how we use the tools and to add new capabilities to help leverage our communication channels. Then soon we lose any sense that we could live without them.

We don't have to rely on the media for opinions: For many years well-known writers or television personalities shared their opinions, editorials, and information through a media of some type: books in the beginning and television later on. This meant that whoever was providing the communication medium was in control of the message, including the perspective or slant on the content. All of this changed when our communication tools went online and the Internet provided ways for all citizens of the world to talk to everyone else in real time, which is something we never had before. Citizen journalism was born.

Citizen journalism is a powerful capability because it provides a way for all of us to hear from any one person in the world instantly and through many different channels. Twitter, blogging, Facebook and Reddit are just a few of the tools that allow any one person to communicate to millions of people instantly and for free. Tools such as these shifted the power of the pen away from traditional media and handed it to everyday people like us if we choose to use it. As the variety of channels of communication grew and as they scaled to be worldwide through the power of the Internet, we could share our editorial opinions faster and farther than ever before.

Once again this is a newfound power that many people do not understand or appreciate. Many folks do not realize how filtered or tainted communications are when they pass through mass media. The evening television news or a newspaper sports writer might deliver an opinion piece that influences thousands of people upon reading or hearing it. Now instead of a dynamic where one media person can impact a large audience, we have thousands of people who can influence one person just as fast. Having billions of people who can report an event or express an opinion to billions of other people completely changes the ways society is influenced. This is no small thing of course. Who we choose to listen to when they communicate has much to do with building our worldview, be it accurate or inaccurate. This is yet another way that technology is altering the world dramatically.

Citizen journalism brings with it much more good than bad. Sure, there are people who post things that are not true or opinions that are a bit twisted. But it is a huge improvement when any event can be reported directly from the actual moment from citizen witnesses. The mainstream media has competition when it comes to influencing public thought, and that is a wonderful dynamic for all of us. As a species now we can communicate instantly and everywhere, and this puts us on the way to toward a Universal Mind that will help each individual person become more enlightened. That is a fantastic thing.

When I ask myself if God created the Internet, concepts like citizen journalism show me evidence that the positives outweigh the negatives.

Companies and individuals can crowdsource to anyone, anywhere in the world: The ability to crowdsource became possible because we have the ability to post any task we need done online and strangers from around the world can compete to do the task. This is a great example of the collateral benefit of having built the capability to communicate without friction. Before the Web and crowdsourcing sites combined to gather online workers in one place, there was not a practical way to ask thousands of people at once what their interest level might be in doing a specific piece of work. In addition we can have the crowd bid on doing the work, which sets up competition so that we create the environment for a high-quality and speedy result. We could do this previously on a small scale through a help-wanted ad in the newspaper or posting something on a piece of paper in a local store, but we did not have the entire world available to us.

Today there are hundreds of Websites that facilitate crowdsourcing in many different ways and for different purposes. We can get a logo made, raise money and get help writing a book. The reality is that we can get just about anything we want done if we use the right Website to communicate with a willing global workforce.

Over the past few years I have advised a company named WeGoLook. com. The owners of this company - Mat and Robin Smith - had a wonderfully simple idea: let's gather thousands of "lookers" from all over the U.S. and match them up with people who want to buy something online. Depending on the kind of purchase to be made you might need someone who is local to go verify that the thing to be purchased is, indeed, as advertised. They essentially crowdsourced lots of people in many cities across the U.S. who are willing to provide this service and then matched them with people or companies that needed local verification. The company has grown wonderfully since that first idea and now serves all kinds of roles that all require some type of verification for an online transaction or business need.

The key to making this all work has been the ability to use a Website, email, and mobile applications to communicate with thousands of lookers and match them with thousands of people with verification needs. The crowd of lookers does not work on salary for the company. The owners have not met the vast majority of the lookers who work for them face-to-face. Their workforce is simply a crowd of workers who have been harnessed to provide a much-needed service. Everyone wins in this scenario. This would not have been possible to do twenty years ago because there was not a communication method to connect thousands of lookers with the specific people that have verification needs.

Frictionless communication amplifies the ability to engage cost-effectively with workers anywhere in the world. Once again this ability to communicate without friction changes the game when it comes to finding workers and having options for hiring. The outcome of crowdsourcing will be spreading prosperity around the world in a more even way. Any part of the world that has plentiful workers and a lower cost base naturally will have an advantage when work is placed. Creativity and the quality of work will improve because the potential workforce that can be tapped is worldwide at a touch of a button. The economic shifts that crowdsourcing will bring will be global in scale.

Frictionless communication has done wonders for the dark side: Although there are lots of positives with improved communication capabilities, there has been an opening of the floodgates for people with bad intentions to ply their trade as well. There are two new dynamics that really concern me. The first is that someone with bad intentions can be anonymous online. They can create any name and profile, existing online under that false cloak. This person could be a powerful hacker, online extortionist, an ex-partner, or a child molester. Although this form of identity dishonesty always has been with us it is even easier to pull off online where people rarely see each other. The bottom line is that it is impossible oftentimes to know that who you are communicating with online actually is the person you believe them to be.

The second concern is that a person with bad intentions can engage with millions of people automatically and instantly, looking for a weak link. They can engage with any one person they choose to target, anywhere in the world. A burglar may not be able to get into our house physically to steal from us, but a hacker can monitor our online communication, learn about us, and then use social engineering to steal from us. All this can be achieved while living thousands of miles from us physically. Hackers can communicate to millions of people through email with a phishing scheme, looking for the hundreds who can be fooled into clicking on a link with a destructive payload.

Anyone with whom we've had a relationship that is angry with us has an ability now to communicate with everyone around us. This means they have the potential to reach our family, our boss, and all of our friends, so with a click of a button, they can embarrass or undermine us. They can do the same by posting embarrassing or cruel information about us to the entire world, for that matter, and we will not be able to stop them from engaging in this form of "badvocacy." We may not even be able to have a site take down content that was communicated through them in most cases. Frictionless communication works both ways: for uplifting someone or for destroying them.

Unfortunately, easier communication created new capabilities for online criminals to amplify anything they can conceive to reach a victim. Technology-augmented communication has become a weapon in their hands and there are many innocent people who are victimized every day.

Exposing young people to everyone, good or bad: When we made communicating with each other easier we also opened up the door for anyone who wants to make a connection to abuse a young person. When a child can see the person they are dealing with face-to-face they have a greater ability to judge the danger in the situation. In most cases they can feel the danger just by being in the dangerous person's presence. But when children communicate online they lose the ability to have all of the information available about the person they are engaging with. Predators are using this dynamic to connect effectively with kids anonymously and directly. Thankfully law enforcement can use this same dynamic to trick predators by acting as if they are children. Unfortunately, law enforcement cannot catch every predator before they reach some of our kids.

Online adult predators are not the only danger that amplified digital communication causes for kids. Easy communication methods allow school bullies, former friends, and hated enemies to attack publicly and privately. Words pointed at us have deep impact, especially for young people. Teasing on the playground now has been replaced by teasing online. Mix this dynamic with the courage that comes often with online anonymity and we have a painful mix of extremely toxic communications going on in the world of many young people. When kids attack electronically often there are not teachers or parents around to overhear and get things under control. The impacts of this form of communication have been devastating in some cases. Kids committing suicide because of what has been communicated about them electronically is about as bad as it gets, and this is becoming an all-too-common outcome.

Young people are a great study in determining the raw value of functionality online. They look at any new communication tool (devices or software) simply as having value to them or not. They are highly motivated to use tools their friends are using and that are low- or no-cost. They look at each new channel of communication and ask, "What unique capabilities will this give me?" If there is something with unique communication tools they will try it out and see how it works for them. Snapchat is a great example of a new form of communication that has the unique property of deleting the content quickly after it is viewed. Because this added a new element to the inventory of ways to communicate, it has taken off very quickly with younger generations.

The problem with some kids' eagerness for experimentation is that they may search out capabilities that will help them avoid oversight or accountability. They are very clever at times about being able to communicate what they want to whom they want, often with a minimum amount of tracking. Again when it is possible for anyone to do activities in the dark, bad things often result. Normally there is a lag time between when young people figure out the next hot application and when their parents catch on. This gap in the speed of adoption often gives younger people a capability for misbehaving because they have command of new tools that facilitate something their parents would not allow if they only understood what was going on. Not doing things that parents would be proud of has been going on forever; the difference is the scale and consequences of behaviors today.

Going forward I hope the digital landscape gets safer for our kids. I hope we learn how to teach them more discretion online and that we find ways to better identify the online criminals before they reach a child. I hope we do a better job of monitoring for bullies and develop ways to dissuade them from using their words, pictures, and videos like clubs to bludgeon young people around them. History has shown that, when we develop new technology concepts, it takes a while to figure out how to govern them properly: not too much, and not too little. It would be tragic if we allow frictionless communication to be something that injures our kids more than it helps them.

Communication can distract us from contemplative thinking: Immersive communication can suck up all of our time, and one of the important aspects of life that can be squeezed out is our time for deep thinking and processing. There are many studies that show the impact of constant interruption caused by our new digital communication channels. My take on what we have found to be true is this:

- We are interrupted at a higher rate today because of the digital intrusions of email, text messaging and social messaging. Couple this with wearable devices that make each communication vibrate on our wrist or chime in our ears and we have a prescription for a constant barrage of attention-grabbing by our technology tools.

- It takes around five minutes to get our mind back to the place that it was before the interruption. If we are interrupted fifty times a day that means we spend 250 minutes a day just trying to get our focus back on something we were doing or thinking about before the distraction.

- In order to maximize our creative talents, we need to be able to focus for extended periods of time on the task at hand. Constant interruptions do not allow our brains to develop innovations or creative solutions to challenges.

Even while writing this book (which is a creative process), I have a constant stream of email notifications, text messages, mobile alerts on my phone, and my wearable device buzzing on my arm to make sure I know about every upcoming appointment. When I lie in bed at night trying to go to sleep, either my phone or my wife's phone is very likely to beep, vibrate, or ring for some reason. I feel the constant pull to respond to my email all weekend because I know if I don't I will be buried under a landslide of important email on Monday morning. I used to be able to get lots of thinking time while I drove for long periods but now people can reach me easily on my mobile device at any time of day or night. I travel constantly, and being on the plane for hours might be the last place where I have a communication-free zone, and even that will go away soon when people are making cell phone calls from the air!

Now, I could put away my smartphone and just go off the grid, but when I do that my family berates me for not being available at a moment's notice. Coworkers and clients also assume they can text me at any moment and get an immediate answer. There is an expectation in my life that I am available to communicate readily. The trade-off for this is lots of interruptions, of course. Immersive communication has lots of upsides, while being able to concentrate for extended periods of time has simply become more challenging.

The Future of Communication
The new tools we use to communicate have become highly integrated parts of our lives as we check our devices to respond to an alert or check social streams over 150 times a day on average. We are surrounded by an abundance of friends, family, coworkers, clients, information providers, vendors, and others, all "talking" to us through a suite of devices and online channels. The difference between the communication capabilities we have now and those available just 35 years ago is monumental. What is even more staggering is that the pace of change is speeding up so that the changes that will come over the next 35 years likely will be hard to imagine today. Where our communication capabilities go in the future is worth speculating about because communicating is one one the most important activities we do as human beings.

A simple place to start is just analyzing how we engage each other with words because we can apply trend extrapolation pretty easily to this simple activity. This is a list of what we have wanted for communicating with words:

- The ability to communicate to one or many inexpensively or for free

- The ability to communicate instantly with anyone, anywhere in the world

- The ability to communicate in various forms of media to hold our words digitally (a Word doc or PDF) or facilitated verbal conversations (Skype or a plain audio connection)

- The ability to choose if the communication requires a fast response or can be answered anytime within a day or two

- The ability to choose how "abstracted" we are when the communication is delivered. Video conferencing is very different than an email. With video calls it is real-time; you will see my face and I will see yours. Email allows me to send a message and not be around when the recipient gets it, and not to deal with any response physically

- The ability for the communication to self-destruct so that it can only be seen once

- The ability to talk to one person or millions with the push of a button. In between those two extremes is the ability to create communities of people of any number we choose

Wearable devices will move the ability to communicate with each other closer to our body, which will give us a more options to receive and send messages in various new forms. We will integrate gesture and voice control into our ability to create messages. We will add new methods to view or hear messages too, such as hearable devices, retinal projection and heads-up display capabilities built into any piece of glass in our eyesight.

Our wearables will help us filter which communications we will need to respond to immediately from those that are meant to be read at some point. A wearable will have an inventory of vibrations, pulses, or sounds that are unique for different people and with different urgency levels. This will provide an alert system that we can customize so that we can monitor a stream of communication all day and receive subtle clues about what we really need to deal with immediately. Our filtering systems also will give us an ability to digest the communications we receive on various devices depending on what is available to us at the moment. We will be able to direct communications with our wearable devices to any size screen with a wave of a hand.

We have no way today of telling the world our status in real time or changing that status to any one of five different states (available, invisible, emergency only, busy, off the grid). This makes it impossible to know if, or how, we can communicate with a person in an instant. Today we are resigned to lob a communication out to a person and hope they answer us quickly. In most cases we have no idea if they will be free in a couple of minutes to connect or if they are overseas on holiday. Posting our status publicly will be a normal activity that everyone does in the future which will change radically how easily we can connect with people instantly.

When we make it easier for people to reach us (and most of us will) there will be a corresponding need to be able to filter incoming communications in a way that is customized to us. What I see coming soon are filtering tools that intelligently help us connect to all the channels of communication we desire. Then we will set rules and parameters that will help us prioritize what we see and react to. We will better prioritize our online relationships so the flow of information we see will be pre-sorted, which will allow us to invest whatever free time we have to communicate more effectively. I have long believed that email as a tool is a crude first-generation application that has not improved at nearly the rate as some of our other inventions. This is the reason that we get too many emails from sources we do not wish to have wasting our time, and too little ability prioritize and triage our email streams so that we can discern the more important conversations. Social technologies have progressed much further and faster than email and they are still young. The ability for us to provide our real-time status to the world so everyone knows the best options for communicating with us at this instant is nonexistent at the moment, but we will have that one day. We have only begun to learn how to use all the communication tools we have been given to flow information. It will look much different ten years from now.

We will continue to improve our ability to "talk" to someone in real time. We will use Brain-Computer Interfaces (BCI) to allow us just to think about a person's name and communicate with them in real time. The concept of technology-aided telepathy has been talked about for years and we may be close to getting there by using a BCI to convert our communicative thoughts into a format that can be transmitted to another person's BCI. At that point we can see the potential of the Universal Mind to allow people to talk to each other regardless of where they are; geography will be a dimension that means nothing to us. We will be moving toward the ability for any one person to connect to many others so that thoughts are shared collectively in real time. Problems will be solved by sharing them instantly with the collective and having anonymous people answer. As far out as that sounds, we already have Websites that provide this capability so the final step is to make it easier to state our issue immediately (at the speed of thought) so that we quickly get solutions.

When it comes to communication, as we get more of what we crave we will continue to evolve as beings because communication is key to sharing information, and information sharing helps us learn and grow. I suppose that is predicated on the quality of the information so let's acknowledge that if we use our powerful new communication capabilities to share worthless information we will not progress. Assuming the majority of us will share valuable communications we will be better informed at escalating rates. We will keep in touch with more people, more often. We will be safer because communication in case of an emergency will be easier, instantaneous and will reach a broader base of people. Introverts will relish going off the grid so they can get away from all of the communications coming at them. We will slowly feel less like individuals and more like integrated parts of humanity no matter where we are in the world and what time of day it is. We will feel very connected to others because we will be able to talk to them effortlessly and the "them" I am referring to will be billions of people.

Some of us will love this feeling while others will push against it, so we should not assume that everyone will want to be connected in an immersive way at all times. Introverts might choose to spend more time off the grid while extroverts will be fed from an energy standpoint by having such a robust connection with the collective. I see a time coming when we will find a balance that is appropriate for our unique situation and live more peacefully by spending parts of our days off the grid so that we can be uninterrupted and have deep thinking time, and other parts immersed in the larger conversation. With that said, our young people today can be very addicted to instant communication. The young people of tomorrow may not be able to discern a difference between being connected or disconnected because they may never choose to be off the grid. For many of them being offline will feel like "death."

SUMMARY:
Frictionless Communication

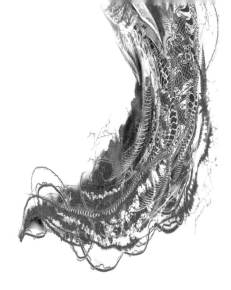

Communicating with each other is one of the most basic human activities. We wither away when this is withheld from us because it is through communication that we connect with other humans. We have been blessed with an explosion of new tools that have improved our ability to communicate with each other manifold. In the space of twenty years we have seen more changes in our capability to communicate with each other than in any other time in history. This newfound capability has enabled completely new dynamics for humanity: Citizen journalism, crowdsourcing, digital marketing, social technologies, and building online relationships.

These new capabilities are having a tremendous impact on the world because they provide an ability to share information from anywhere, to anywhere, instantly. That changes lives because it helps shine a light on dark behavior. It helps us know more about each other and update our status more frequently and for more people. The world is becoming a bit safer, more informed, and closer together because we can communicate much better than in decades past. Clear lines of communication have always helped us be closer even if we have very different beliefs.

My light-versus-dark ratio with frictionless communication capabilities is 60/40 toward good. There is more to like about what this is bringing us than there is to fear. To be sure, there are serious dangers that I mentioned earlier. Offsetting these are a number of positives that we are growing numb to already and which are very substantial. We must not lose sight of the positives of citizen journalism, social technologies, online connections, and the ability to have a voice that can talk to billions of people for free.

This awareness of each other through greater communication generally will have a positive impact on humanity. Evil hides in darkness and because goodness is light, we will find that the more we know about the people around us, the more we will evolve. There will be less of an ability to hide what we are doing which will result in a more distinct line between someone who is living a healthy, positive lifestyle and someone who wants to hide their behavior in the shadows. Yes, people can deceive those around them with what they post online by only posting the good things or even lying about what is going on. People have been deceiving themselves and each other forever. What we are finding is that our new communication technologies have a way of shining a light on our behavior because either we confess what we are doing with others online through our social postings or because governments and people around us post information about our behavior through public forums.

Humalogy Viewpoint

Over the past century or so the only electronic technologies we used to help us communicate were the telephone, radio and television. These certainly altered the course of humanity in the 20th century. The 21st century will be impacted by Web-based tools that have dramatically improved our options for communicating with a single person, groups, and the world in general. At each level we can communicate now in words, videos, and pictures. We can do this for a very small amount of money, and when we do wish to communicate the delivery is instant regardless of distance. For this reason we have seen the Humalogical Balance shift toward the T side.

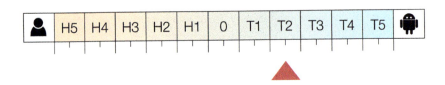

I place our overall use of technology in communicating at a T2. For this measurement I am taking into account the volume of email, text messaging and social technologies that we use now to talk to each other. I also believe that this is not a bad balance point if efficiency, distance, and volume are factored in. What I mean by this is that we could be at H5 if we only talked to people around us. The problem with that is that we would not be talking to anyone who is not in our presence or to larger groups of disparate listeners.

That limits the amount of people we would be able to communicate with drastically. One could argue that we should be at a 0 on the scale in order to have balance between in-person conversations that have deeper meaning and remote conversations that will feel less personal. I disagree with this because I believe there is great value in the scalability and ability to talk to people who are not in our presence. Email is a great example of this because it lets me "talk" to over a hundred people a day. If we go too far in the other direction and have the vast majority of our communication done through technology and do not engage with people face-to-face, we will lose the human connection that is so important for our wellbeing. For this reason, T2 might be the best balance point for communication.

Seminal Questions

Will our worldwide ability to communicate without friction help spread a higher quality of life more broadly across the world?

Will our improving ability to communicate help us learn faster and become more enlightened through the ability to connect with many more people?

Ultimately, will we become comfortable with an unfettered ability to communicate instantly with millions of people around the world at the speed of thought?

Will we awaken to how our words affect the whole forever and take greater care in how we communicate, or will we continue to have online trolls who invest their time in injuring others?

At what point in the improvement of our communication technologies will we have a real-time Universal Mind?

CHAPTER SEVEN:
Privacy, Transparency, and the New Normal

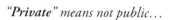

*"**Private**" means not public...*
*"**Secret**" is an intention to keep knowledge from others...*
*"**Transparent**" means visible to the eye or*
completely public...

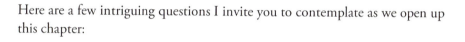

Here are a few intriguing questions I invite you to contemplate as we open up this chapter:

- How much of a right to privacy do we deserve as citizens of this world?

- Where does this right come from? If your answer is from some form of government then you probably understand that this is why we have so many different definitions of what is private across the world.

- Who should really be choosing what is private and what is not: you, me, or a government?

- Is privacy really a concept that is outdated at this point?

- Would we all be better off in a world of complete transparency so what happens in the dark can be brought into to the light?

The very concept of privacy is much more complicated than the cultural meaning of the word. For us to understand what technology is doing to our privacy, we need to contemplate the idea of privacy in some detail. Most people who are going to behave badly use the cloak of privacy to assure that society does not "see" what they are doing. So logically, we could improve the world by removing most of the privacy we now enjoy. The level of privacy or transparency that we should have as a human right is an interesting question, and it is highly impacted by technology. We are moving into a time where this will become a huge debate because we continue to invent and deploy technology that is reducing our levels of privacy, whether we agree with that or not.

On the other hand, without some level of privacy, governments and dictators can punish anyone they choose when that person does not see the world in a way the overlords wish. It would be virtually impossible to have healthy insurrection against bad leaders without an ability to hide one's true beliefs and actions. Privacy allows for individualism because it provides a cloak so that a person can be themselves and live their own truth without the danger of the people around them pulling them back into the herd. This is such a difficult issue because, behind that cloak, there can be fruitful, as well as devastating, behaviors going on.

Another healthy reason for privacy is to provide basic safety from people with criminal intent. The less someone with bad intentions can find out about us, the safer we are. As a person acquires more wealth or creates enemies in some way, the more they have a need to restrict public information about where they live, how they travel, and details about their family members. If there was no evil in the world there would be much less need for privacy. This points out interesting spiritual principals that weave into the discussion of privacy because, in a perfectly enlightened spiritual world, there would be almost no need for privacy. We would not be participating in behaviors we needed to hide and we would not have people doing evil things that we needed to hide from. Depending on what you believe you might capitulate to a reality where privacy is needed always because we never become very enlightened as a species, or you might be more hopeful and see less of a need for privacy over time.

My observation is that, just in my lifetime, I have seen a shift in the desire for privacy. People are less modest from a physical privacy standpoint and more willing to show parts of their body, whether they look good or not. People are willing to share lots of details about their personal lives through the use of social technologies. We build public profiles on sites like LinkedIn so that anyone can find us and learn a little bit about our career information. People are more comfortable in verbal discussions sharing with others the issues we all face, even when embarrassing. We seem to have a growing maturity that we all have body parts and we all have issues in our lives, and it is okay to acknowledge these publicly. At least this is true in the

U.S. as we are becoming much less Puritanical about showing vast real estate of flesh. At times, a little too willing to ignore the concept of modesty! In the land of the free we are slowly working our way through what we should be free to keep to ourselves and what the public or government need to know.

A difficult balance that we all face is between the concepts of privacy, freedom, and figuring out if we can have one without the other. Freedom is a critical tenet in the U.S. because it was one of the foundations this country was built on. We have the freedom to do as we choose as long as it stays within the boundaries of the law and does not impact others around us negatively, for the most part. We have freedom of speech, freedom to choose what we do with out time, religious freedom, and the freedom to succeed nearly without limit or fail economically. We have also had the freedom to be as private as we wanted with our behaviors. Today we bristle at the thought of the government monitoring what we do, listening to our phone calls and conversations or tracking our whereabouts. We see privacy as a key component of the freedom we so enjoy.

Are freedom and privacy really conjoined? Is it not possible to be free to choose to do anything we like and, at the same time, have everything we do monitored and archived? I am not suggesting this by the way; I do want to point out, however, that conceptually we can have total freedom to do whatever we like and, at the same time, have little privacy. No one would stop us from doing what we want (within the bounds of the law) yet everyone would be able to track and see what we choose to do. This would not apply necessary to what we do in the bathroom or bedroom, but certainly everywhere else. For most of us there is a knee-jerk reaction that is not positive to this concept.

Let me pose another question: would you trade away nearly all your privacy in return for a huge amount of safety and security? Truth to tell, I would be fine with this bargain. I feel I have nothing to hide so I don't care if someone monitors all my travel, all of my activities, and communications if that means they were doing the same with everyone else and could, thereby, protect me from danger. The problem with this of course is the "who" would do the monitoring of everything in my life: a government, a private company? I would need to trust that they would only use the information for good and not to coerce me in any way.

So do I trust the U.S. government and Google to know everything about me? Here is my honest answer and I know some people might not agree with me. I trust the government because I really have nothing to hide from them; I pay my taxes, I follow the rules and I mind the laws, so if they want to know everything about me and everyone around me, and in exchange that prevents acts of domestic terrorism, I am good with that. At the end of the day I trust them at a fairly high level because their motive generally is to protect citizens and freedom. I am not naïve in that I know that governments might not be able to be trusted 100% and, given enough personal information, they might overstep moral boundaries 9in in some cases they have). I also understand that individuals in the government might go rogue and misuse my information. With that said the world of privacy is a world of trade-offs and choosing the lesser of two dangers at times. When it comes to protecting our homeland or protecting my personal privacy, I will lean toward protecting our nation.

It is a different story with Google. They have a very different motive in that they are built to make a profit. To do that they will sell my information to anyone they can without causing a public furor. They will do all they can not to let us know, too, how much they know about us or what they do with our information. Their well-publicized mantra of "do no evil" is noble and I like the thought. I just have a hard time trusting a for-profit company where my data (private and otherwise) is used as the currency. In fairness, they offer many free services and the bargain we make to use them is sharing our data with them. My concern is that most people do not really understand what they are paying for the "free" services. I am only using Google as an example of a for-profit company that trades free services for data; Facebook, LinkedIn, and just about every other social site does the same.

If you do not have a clear picture of why I say we are already advancing rapidly toward having significantly less privacy, think about this; Governments are putting cameras in many public places, adding more all the time. Telecom companies have records of all of our calls and, in some cases, where our phone has been and therefore where we have been. Social sites gather lots of data on what we post, who our friends are, and where we have been. Search engine companies know everything we search for and sometimes where we are surfing on the Web. Marketers are putting together what we purchase with any other piece of data on us they can grab. Our mobile applications, if so enabled, can track loads of information about where we go and what we do. Add all of this up and we see that carrying a mobile phone, doing online transactions, and being out in public can add up to being fairly transparent with our lives, whether we realize this or not.

Just to give you an example of how arbitrary we are with respect to deciding what we want for privacy, we passed HIPAA laws in healthcare so that the industry cannot share our individual medical data with anyone. This helps keep anything that is happening to us that involves a physician private for the most part. At the same time a retailer can grab the data on everything we buy and sell it to anyone they want, thereby making money from our personal purchase data. Think about some of the things you purchase and then consider how you would feel if that information was made public to everyone. In fact we are not that far from this reality. Banks sell our purchase data that they gather from our use of debit and credit cards. Once again they make money from our personal information on purchases and do not disclose this to us by and large or even feel bad about it. Of course they do this very quietly and most customers have no idea. Buried deep in the small print of their online banking applications often is a disclaimer that they can provide our purchase data to others.

There is a great story in the Bible about Adam and Eve being naked. Before Adam and Eve ate from the fruit of the forbidden tree of knowledge, they ran around the Garden of Eden without clothes and it did not bother them at all. Once the serpent had a conversation with Eve and convinced her it was okay to eat the fruit, they took an action that they knew for sure was forbidden by God. This action introduced sin into their lives and the shame that came with it. This perceived shame is what caused them to knit together aprons of leaves and cover themselves to hide their nakedness. So in the beginning we had no sin and hence, no shame. This story illustrates one of the reasons that someone might want privacy; they believe they have something to hide and do not want it exposed to the world.

Body modesty is not our only desire for privacy of course; there are logical reasons to restrict personal information from the public as well, like the fact that we are on vacation. What is important to discern is that not all reasons for privacy should dictate that everyone should have a full right to privacy. Full rights to privacy sound great in theory but, in a civilized society, there are some actions and behaviors that need to have a light shined on them.

Let's dissect privacy into a few different parts in order to really see what is happening to us today in this area.

Body privacy: Physical modesty is one reason we wear clothes, and to keep warm is another. Most people are not predisposed to running around naked. There are really good reasons for physical modesty and these differ depending on the stage of life we are in. When we talk about babies it is acceptable for close family members to see the child naked while changing diapers or perhaps during a bath. As a person grows up it becomes more necessary to keep certain parts of our bodies covered because these are looked at through a sexual lens. Being animals at our core, we are attracted to these parts of the body, sometimes in very unhealthy ways. As we get older we may not like how our bodies look. We are very motivated to disguise or hide our bodies because we judge ourselves and are afraid that others will judge us. More important, when adults show excessive amounts of skin to other adults it sends signals - intended or not - that might cause unwanted attention. People posting nude pictures of other people without their permission is a good way to be thrown in jail. Trying to take pictures with our cell phone up a girl's skirt will land us in jail. Even without technology a peeping tom looking in a window for the express purpose of seeing someone naked goes to jail. We take this form of privacy very seriously when compared to other forms.

Body privacy technology problems: Technology is giving rise to all kinds of dilemmas in this area. Cameras in public places are recording increasingly more of our actions and, in most cases, we have no idea that these cameras are recording us to video. This means that if someone is able to put a camera in a bathroom, there will be massive amounts of privacy invasion. Hackers and certain software programs can take control of the camera on a laptop or mobile device and provide them with access to whatever our device is looking at, including us with our clothes off. Just the other day a lady stopped me at a speech I was giving and railed at me with the following complaint: She said she had purchased online coaching from a firm and the software the coach was using to work with her remotely turned on her camera so they could communicate face-to-face. The only problem was that the coach could turn on her camera anytime he wanted without her permission. She had just realized that at times she would work on her computer in the buff. You can see the problem that concerned her easily enough.

Add to this the fact that more devices are being sold with tiny cameras built in - including all of our smartphones now and you have a prescription for people having the ability to take pictures in all kinds of places where someone might not be fully dressed. To sum up, we have much less control today over the world seeing parts of our bodies than we once did.

Activity privacy: This type of privacy may be the one that gives people the most heartburn. Answer this question: how would you feel if every single thing you did in private was broadcast out so the entire world could see? Would there be anything illegal? Embarrassing? Guilt-inducing? Heartbreaking? For the vast majority of the world this answer is "yes." Even with our clothes on many of us would not relish the world knowing every move we make as we make them. We all do things when we are alone or when we are with certain people that we don't want others, or anyone, to know about. There are bodily functions, addictions, affairs, habits, places we go and many halfway normal activities that we would not want anyone to know about. Some people believe that no one has a right to know anything they do, even if all of their activities are harmless. There are others who are deep into many activities they know are wrong, and their reason for wanting privacy has more to do with knowing that being discovered would lead to grave personal repercussions.

Each of us maintains a very complicated list of activities that we want either known or hidden from various people in our lives. Some people can know some things and others need to be kept in the dark. Activities might need to be hidden from authorities while we consider it okay for a friend to know. Some people believe their spouse might know a lot but there are still a few things that are better kept hidden just to keep peace. I believe we all carry a privacy belief matrix around our activities that is unique to each of us.

Activity privacy problems: Technology provides many ways to record or extrapolate what we are doing. If we have conversations using our devices other people can track them. If anyone takes pictures of us doing anything this activity is memorialized. That can be smoking pot, having sex, picking our nose or driving through a red light. Our phone providers know everyone we call or text and even the content of what we say (or show) to each other over our devices. In order to keep us safer the government is putting cameras up in public places. They are recording who drives down a road, if they ran a red light and even who was in the car. Since we are still in the early days with all of our new Web-based technologies, there are still many places outside our homes that do not have video coverage, but how long will it be before they have cameras up in a wide network so they can capture just about any activity outside our homes?

This lack of privacy outside the walls of our homes gets even worse if we start thinking about how easy it is to fly a drone with a camera over our property in order to film whatever we are doing outside, or maybe even looking in our window. The new peeping Tom might have four rotors and a tiny camera recording our activities within the privacy of our own homes.

Finally, we have the personal recording of our activities from wearable devices that have tiny cameras built in. Police forces are moving more in this direction as a way to better memorialize what really happened when they engage in situations. We will likely see this broaden out to the general public who will look to record on demand when a situation calls for it. Google Glass caused a furor when it was introduced because it allowed the wearer to record video or pictures with a simple verbal command or touch of a button. I am sure this is just the first in a series of devices that will allow a person to shoot video easily without having to buy special surveillance gear. At least when someone holds up their smartphone or tablet we know what they are doing. Wearable devices will take that visual cue away because they will be able to record more discreetly. Very soon we will have to assume that, while out in public, we will be recorded every moment by a static camera on a pole or moving cameras on the people around us.

The days are nearly gone of being able to do activities anonymously or to behave badly in public and get away without consequences. Although this may be perceived as a huge negative change in our lives, we are already learning that when people understand they can be videoed while behaving badly they behave better in general. This has been evidenced with police, angry customers, or rude service providers.

Information privacy: There are thousands of fields of data now that can describe each one of us. I'm not speaking about what we are doing; that was covered earlier. I'm referring to demographic, psychographic, physical, or statistical data about a person. In the past someone around us would only be able to access the information that we told them verbally or that we would be willing to commit to a piece of paper. Even then this data was restricted to a file cabinet somewhere. This meant that a person who wanted privacy could guarantee it pretty much by refusing to share any information about themselves. Today that is not practical because it would require a person not to interface with government, healthcare, the education system, or any technology at all. It is possible to keep ourselves off the grid and away from databases but we would have to hide very deep into the woods and never come out.

For the rest of us there is a new reality about information privacy; we have very little control over what people gather and share about us. If it were possible to see all the fields of information that every organization in the world is tracking on us I am sure we would be surprised. Moreover, if we could see how easily that information flows between organizations we would be shocked. Databases are everywhere these days and we are filling them with information about each other.

When I was about eight years old I was talking to my Grandma Baker. She was about 4 ft. 11in.: not a big woman. I was a curious child so I was asking her questions about her job at a museum, and she was fine telling me what she did and what her title was. The discussion was going along fine until I asked her how much money she made a year. At that point she slapped me across the face. I was shocked of course and she proceeded to tell me that you never ask a person how much they make in salary because that is private information. I had asked my mom that question and she gave me an answer, not thinking anything of it. Grandma Baker went on to tell me there were many things that just were not my business. This was one of the first experiences I had with unknowingly crossing over a privacy line. I learned soon thereafter that people had unique viewpoints on how much data about themselves was appropriate to share with me. I was also taught that data about myself needed to be withheld carefully from strangers because it could be dangerous for them to know where I lived or who my parents were. In fact the less anyone knew about me, the safer I would be. My Grandma grew up through the Depression and it was a different time. In her world, there was safety in anonymity. My children are growing up within a very different model for sharing information.

Where we once had tight control over our personal information, now we have begun to lose any control. In fact often we have no idea who knows what about us or with whom they are sharing our fields of data.

Information privacy problems: Doxing is a word you may not be familiar with:

Doxing (from dox, abbreviation of documents), or doxxing, is the Internet-based practice of researching and broadcasting personally identifiable information about an individual. The methods employed to acquire this information include searching publicly available databases and social media websites (like Facebook), hacking, and social engineering. It is closely related to cyber-vigilantism and hacktivism.

Source: Wikipedia.com

The fact that we have a new term to describe broadcasting someone's private information publicly is a sign of the times.

In our zeal to automate the world with computers and software we have gotten rid of paper-based information storage for the most part. This created an ability to flow data between companies very easily. It led to data being looked at as a sellable commodity at worst, and a shareable asset at the least. Unless it is regulated by the government like our HIPAA laws, companies who gather data on us knowingly and unknowingly have all the motivation in the world to use, sell and trade our data with anyone willing to pay. In the U.S. we have virtually no control over this. Just as a point of reference, in Germany the privacy laws are much tighter and people do have more control over their data. This is the reason that Google was forced to provide a way in Europe for people to have information deleted from Google if they choose.

Most of this is happening in the background of our lives so we have little sense as to how well we are "known" by any one entity. Technology companies are walking a thin line because they are trying to be as open as they can be with the government so they are not regulated. At the same time they really do not want the general public to understand how much they track on us. I read a story once about the Facebook data scientist team and the point of the article was that they have to bleed out their analytics and measurements slowly to the public because if they turned loose of everything they were learning about users there would be a privacy invasion outrage.

As I mentioned earlier companies in the social technology space and the search engine industry use our data as a product to sell to advertisers. They give us their services for free because they want everyone to use their tools. This allows them to gather more fields of data on us. The more they gather, the more they can charge advertisers. We agree tacitly to this deal because we like their free tools, and most people have no idea what they are trading away. There does seem to be a growing inkling about the depth of information these companies inventory on each of us now. Unless regulated by a government – and I am not advocating this, by the way - the for-profit companies will continue to build ever bigger databases of information about us all. If we are using the Web they know something about us. If we use online tools they know us even better. If we communicate with friends through their services and buy online they know a lot about us.

It is important to note that there area growing number of organizations capturing data about us without our permission. I mentioned earlier examples like the items we buy at a store being linked to us and banks capturing every transaction we run on our debit cards. Going further, note that a credit card company captures every purchase made on their card. A smartphone or mobile application has the potential to track everywhere we go geographically and when we we go there. Roadside assistance technologies like OnStar in our new cars tracks everywhere we have been, how fast we have been driving and a few other choice pieces of data. In some cases this data is tracked whether we have the services turned on or not. This list grows much longer of course, and in many of these cases our data is being sold or rented to advertisers and data aggregators without our knowledge.

Technologies' Impact on Privacy: Benefit or Disaster?

Now that you have this perspective on the three different types of privacy - Body, Activity and Information - now we can discuss more concisely what is happening with privacy in our lives. Then can take a look at whether this is a good thing for humanity or a growing disaster. One reason this is a really critical issue is because it is going to be hard to go backward once information, pictures, video and action logs are created and stored on each of us. It is very unlikely that companies will destroy it all, even under penalty of law. Either we need to be at peace with much less privacy or we need to put in place much harsher and more detailed regulations now.

One might wonder if younger generations will have less of a concern over privacy forever or if that will change as they get older and more jaded about the world. Maybe as they get older they will feel like they have more to protect and will get more conservative about sharing online. My belief is that there will be a constant change in behavior over the next few generations, with each one refining the "rules" around what is considered private and what is accepted as okay to be public. At some point we may come to a more generalized acceptance of how much privacy we should expect, and it will certainly be much less than we have had in the past. Often with such shifts we go from one end of the spectrum to the other, then swing back to the middle point. So we should be prepared over the next two decades to have progressively less privacy, followed by a shift back to having an agreed-upon set of information that is kept from the public market.

If you are shocked or anxious about my observation that we will have much less privacy over the next twenty years, let me unpack that a bit for you. Organizations have begun recently to buy into the concept of data is a powerful raw material. We have had the ability to store electronic data for a long time. The amount we can generate, store, analyze and manipulate now is staggering compared even to ten years ago. As data becomes less expensive to own and easier to acquire, we are becoming fascinated with all that the data can tell us when analyzed. The era of Big Data is defined by an excitement about having visibility into trends, anomalies, patterns, and analyses we simply have never had the ability to see until now. Facebook is able to find patterns in our behavior because they can aggregate data across more than a billion users and in some cases they are able to note meaningful truths that we have not understood before. For example, because they track when people get into, and out of, relationships, they are able to graph the most likely times of the week, month, or year that people break up. The National Security Agency (NSA) uses Big Data to find connections and patterns concerning possible terrorist activities that would have been invisible to us in the past. Although some people bristle at their use of OUR collective personal data to do this, not many are complaining about the attacks they avoid for us. We will get back to this debate in a moment.

As we fall more in love with the power of data we hunger to build our databases even larger. In many cases the databases we are building contain information on each other. Digital marketing is predicated on collecting data about the people an organization wants to market to and there are many organizations trying to market to all of us. That makes for a lot of entities with a greed to learn more about us, whether we consider that information private or not.

At this point you might be wondering who it will be that will step up to rein in the free-for-all that is data gathering at the moment. Will it be corporations who come under enough public and financial pressure that they are forced to instill a higher level of privacy for people? I am skeptical about this because putting any limits on gathering or using data just takes away potential revenue sources or capabilities. Yes, they will make an effort and try to look like they care about privacy, but it is mostly just for show so that they can continue to keep governments out of their hair on this issue.

The more likely source of progress with privacy will come from governments. Because they represent their citizens they will put laws in place as the cry for privacy hits a high enough level, just as it did with healthcare data and the HIPAA laws. This will vary from by country and state, and will reflect the constituents' collective beliefs about what is private. Cultural differences will play a role just like they do today with Germany being one of the harsher defenders of data privacy. As citizens learn how much their privacy is being invaded by corporations they will scream and governments will react. What they will not do is slow down their march to gather data for their purposes because even if citizens scream governments can claim that they invade privacy for the greater good. They will be fine developing regulations like the HIPPA laws in health care but will be glad to record phone call data without our permission.

The revelations that Edward Snowden, the computer expert, former CIA employee and government NSA released in 2013 were a shock to many in the U.S. We could not conceive how deeply our government was willing to go into our private lives in order to protect us from danger. They were using tactics that arguably were illegal in order to slow terrorism. There is, to this day, a pretty even split between those who believe that Snowden is a patriot for outing the NSA and those who see him as a traitor for weakening our ability to use data to protect our borders. The response to Snowden shining a light on the government's tactics has been interesting.

34% of those who are aware of the surveillance programs (30% of all adults) have taken at least one step to hide or shield their information from the government.

25% of those who are aware of the surveillance programs (22% of all adults) say they have changed the patterns of their own use of various technological platforms "a great deal" or "somewhat" since the Snowden revelations.

Source: Pewinternet.org

People seem to be either thankful for his courage to let the public know their private communications are being monitored and see him as a patriot, or they are livid that he broke his commitments to the government to keep secrets and they see him as a traitor. There is another group of people who don't care much one way or the other because they don't care what the government knows about them.

I do believe that at some point state and federal governments will intervene and drag us back to some level of privacy even though, at the same time, they may be the worst abusers of privacy. They will rationalize this dichotomy by arguing that it is done for the greater good.

Maybe one way to help us all find peace around this issue is to move the conversation to the following questions: What is the overall usefulness of privacy? And, how will having less privacy change us as individuals? Both of these are open to large debate and in order to really define and regulate where we go in the future with technology and privacy, these must be addressed. You might take a moment an ponder your answers before I give you my view.

What is the Overall Usefulness of Privacy?

The primary reason we do not want information shared about us is that we fear the consequences of what might happen if the world, a group, or a specific person were to acquire data about us. In other words, what you don't know about me cannot hurt me. It really is that simple. From this basic truth the rest of the reason is that there are healthy and unhealthy reasons for wanting information withheld about us. A healthy reason would be that by allowing information about us to be in the wrong hands, it could expose us to someone with evil intentions like a burglar who wants to know when we are on vacation so they can rob the house, for example. If information about our kids is exposed online, we risk their safety. If information about our wealth is publicized, we might attract criminals who want to take what we have. There are some pretty sinister things going on today with organized criminals using seemingly innocuous data to perpetrate a crime.

Here is a good reason for our data to be hidden. Our goddaughter registered for college and got a random email that looked pretty official containing a lot of information about what she would be doing at college as an incoming freshman. It turns out the email was a scam from a group that aimed to lure incoming students into the sex trade. When things like this happen it makes an impression that our personal information is too freely available. There are many unhealthy reasons to hide information which normally involve a person is doing something they know is wrong. This could be having an affair, struggling with an addiction, or abusing someone close to them.

In both cases a person wants privacy while the moral justification behind the two different ends of the spectrum is quite different. There also is plenty of gray area in this spectrum. I might want privacy because the data someone would get about me is embarrassing to me. The information might not show that I am doing anything immoral or illegal and it may not be used against me. But I might feel violated and embarrassed just because of others knowing something "private" that I wanted hidden, like information that I have six toes or a third nipple. This is nothing earth-shattering but I might not want the world to know it. I don't have either one of these just to put your curiosity to rest.

This spectrum of reasons for people to want privacy is why I respond often to questions about privacy by asking the person: "What are you trying to hide?" I don't mean that in a negative sense; I am just trying to figure out how to answer their questions about privacy with as much insight as possible. If they are trying to hide personal information for their own security or safety (where they live, their phone number, their social security number), I tell them I think that kind of information should be kept private for sure. If they say to me, "What I do is nobody else's business and I don't want people knowing anything about me," I have to make the assumption that they might be trying to hide something they are not proud of. While my guess might be right or wrong, people who spend their lives standing in the light and running their lives with high moral and ethical standards often are not overly concerned about privacy. The only data they care to keep private is information that provides basic security. People who are doing things they know are wrong usually are the first to be concerned that their digital privacy be invaded.

How Will Having Less Privacy Change Us As Human Beings?

Depending on your behavior in this world, it's most likely you will answer this question very differently. If you want privacy because you are doing things that you know are not healthy you will surmise that the world will be a terrible place unless you can hide your bad behavior (and receive no consequences). If you only want privacy because you don't want to open yourself to security or criminal issues you will likely believe that the world would be a much better place if there is a high degree of transparency.

Because now there will be a constant challenge to keep anything private we will adjust by making a conscious decision about what we really care about being private. We will cling to a precious few fields of data that we will reserve to ourselves while succumbing to the reality that much of what we do will be transparent. We will capitulate in most cases to the fact that we do not have much privacy so based on our unique personal situation we will draw a line somewhere and seek to control the most dangerous information that could get out.

Depending on our status in the world, how much we have to lose or whom we fear, the list of what will be acceptable to be public will change, and rightly so. A lawyer with lots of assets, a family with kids and a practice in which he or she might anger people regularly might want to be as anonymous as possible. A salesperson who makes a living influencing people and networking with the public to find prospects might want a very public profile. How we exist in the world, our age, wealth or occupation all will be factors in deciding how much transparency vs. privacy we might want. I believe we will find a Privacy Balance Point that is appropriate for each of us and actually be able to enforce it at some point.

In order to provide an example of how we might find our Privacy Balance Point going forward, let me use myself as an example. Note that my concern about privacy is less than most people, so I am just giving my personal view. I will break down a few of the more basic pieces of *informational data* and give you my opinion about what I am fine with others knowing. The first classification of information is descriptive: fields like male, 53 years old, Caucasian, lives in Oklahoma, grew up in Cleveland, played piano, likes Indian food, vacations at beaches and in the mountains, snowboards, coaches girls' soccer, and the list goes deeper than this of course. In this case I don't really care if the public knows any of this about me. I am not in a situation where anything I have ever done would hurt my career or relationships. I have not been arrested or really done anything that gives people a reason to distrust me.

While the list above is positive in general, there are descriptive pieces of data that are less flattering. I bankrupted a company I once owned and sometimes investors in deals I have done have lost money. I have had a couple of consulting clients who were mad at me for perceived poor performance. I have been divorced and certainly some of the women I dated earlier in my life would say negative things about me. All in all, I am fine if the world knows any of this about me. These are examples of the fields of data that for me are on the side of the balance point I am fine being public.

There are additional fields of data that would describe activity data like that I left the house at 7:30 AM, ate at Panera Bread at 11:38, drove my truck and exceeded the speed limit by and average of 5 mph. Again I am not worried if people know where I am and what I am doing. I don't do things I would want to hide from anyone so it is fine if people want to track me. Plus, I don't really have bitter enemies so I don't get too concerned about someone tracking me down. This is certainly not the case for people who have jobs like police officers, politicians, movie stars, and other high-profile or controversial careers. They have great reasons for not wanting to be tracked. There are activities from the past that are really unfavorable for me: times I got drunk and acted the fool, times when I was a kid that I shoplifted and times the police brought me home for mischief. Again, I don't care if the world knows these things because they were long ago and I have improved, and if someone asked me about failings I had, I would be honest and say I was being stupid back then.

The final classification of data is the *body information* and that would include pictures and video of me with my clothes ON of course. This kind of information can go public online because I created it and posted it online or because others captured me digitally, including the government entities who capture images of me just about anywhere outside of my home. In reality I cannot stop the government from tracking me if they so choose. I also give speeches for a living on the speakers' circuit so people video me and take pictures all the time. I assume that I can be recorded anytime, anywhere outside of my home. That includes things that I say so I am careful never to say anything I would not be happy having go public. That means I do not swear or use any type of derogatory language, ever. If someone records a conversation and publishes it without my permission I might be miffed, but I would never be embarrassed. I may not like it but I assume anything I say or do outside my house is possibly public. Generally that is okay with me.

Some of you reading this will relate to how I feel and some will think it is crazy to be so transparent. My point is that we will have an increasing ability to control some of these things. Others, like the government videoing us in public, we will not be able to control. We will have a limited ability to guard our privacy and each of us will draw the line where we choose. Today I meet people who refuse to use social sites because they have no interest in other people knowing about them. I meet others whose lives are open books. There will be no hard and fast rules for privacy and transparency in the future; each of us will draw the line where we are comfortable based on our needs or concerns. We all will find our Privacy Balance Point and will have better tools and laws to do this as time goes on.

In the future the Primary Balance Point will lie along a continuum. At one end will be the "Transparents," who are the people who couldn't care less what anyone knows about them and who will take no action to hide anything. At the other end will be the "Disconnects," who don't want anyone to know anything about them and who will go to great lengths to stay off the grid and out of visible range of the government or companies that want to know anything about them. Although there will be people at both ends of the spectrum there will be a bell curve with many people falling somewhere in the middle. I expect this bell curve to shift slowly toward the Transparents because it will be harder to live at the disconnected end and still have any quality of life. In fact people who disconnect completely will be shunned at some level on the suspicion that they have something to hide. Even today the person who works hard to go off the grid is looked at as being somewhat outside of society. I think this is unfair in that quality of life can actually be better for someone who is off the grid depending on their life choices.

Without being a bigot against those who choose to disconnect (some for healthy reasons, and others with something to hide) I see society shifting toward transparency as a good thing for the world in general. I do believe that living in the light is more uplifting and honest than seeking to be in the darkness. I recognize that some people have made mistakes they are not proud of, and they would like those struck from the public record. I told you that I have a list of mistakes as well. The lesser of two dangers is to have a world where we have a high degree of transparency. The cost of that is much lower than the danger of people being able to hide big parts of themselves from unsuspecting people who might connect with people hiding dangerous tendencies. For this reason, I believe that we will ultimately end up with the balance point on privacy shifting greatly over the next few generations. Sorry Grandma!

In a spiritual sense we were created to be transparent with no ability to hide our behavior from those around us. I believe our souls have no interest in hiding who we are from each other. Our souls seek to deal with the Truth of things. Seeing Truth is painful to us sometimes because it shows us the misery we might be causing ourselves or others. The Truth is what sets us free because it allows us to deal with the reality of life. The Truth is that less privacy and more transparency ensures that we have a clearer view of the truth. We need to learn how to discern the difference between the kind of privacy that makes sense from a safety and security standpoint and privacy used to hide wickedness or weakness.

People will debate endlessly about what God would see as acceptable. My purpose is not to draw those distinctions here. I want to point out that technology is creating transparency and, over time, I think it will shift all of humanity toward a more transparent world. I see this as a slow shift from being small beings and toward becoming greater. For every story of our privacy being invaded by hackers or enemies there will be hundreds of times that transparency will provide either a deterrent or consequences that will be positive.

Someday everyone in the world may choose to follow the Golden Rule, and at that point privacy will cease to be such an issue. Transparency will be much easier to accept when people have little to hide and criminals have no way to be anonymous. The technology tools that we are inventing and already have developed give us amazing abilities to capture and store data about people. There is no way we will roll back the clock to a time when people could remain completely anonymous if they chose. This information will be stored for a very long time so our lives will be documented by many organizations in many ways, and the compilation of all this information will paint a detailed picture of our lives.

Companies who aggregate online data will abuse their capabilities and will have to deal with the rules of law that will spring out of their overreaching to make profits.

Governments will overstep as well, and where they have not earned trust, they will have much push-back from citizens. People will snoop on their spouses, friends, kids, and family members. There will be millions of examples of people abusing technology to infringe on privacy for negative reasons. This will invade our civil liberties for sure and will infuriate many people (and rightly so). All of this will help us find the Privacy Balance Point that seems right for society.

SUMMARY:
Privacy, Transparency, and the New Normal

There are healthy and unhealthy reasons to want privacy and organizations will allow people with healthy reasons to maintain some semblance of this if they want it. When it comes to criminals, abusers, bullies and jerks, there will be more light shined on what they do. This will help society provide swifter consequences and hopefully will act more as a deterrent over time. Fewer people would perpetrate crimes if they absolutely knew that they would be identified right away. Through a spiritual lens this is a wonderful outcome. Certainly God wants fewer people perpetrating evil in the world, so anything technology can do to shine a light on destructive behavior will be seen as positive by most of society. By shining a light with transparency the goal is not to put a huge population in jail or for us to punish everyone who does something untoward. The outcome we want is the avoidance of unhealthy behaviors because people know they will be found out and will have consequences to face.

With that said, over time technology-driven transparency will result in a more civilized society because it will shine a light on bad behavior and help reduce the amount of crime and violence in the world today. This will happen – and already is happening - on a country level as well, as any harsh behavior by a government will be posted online immediately. On an individual level knowing that our family can track our behavior in a much closer way will encourage us to stay on a more positive track.

Concerning the shifting of privacy and transparency I see this as 80% positive for humanity and 20% negative. There will be a lot of people who disagree because the concept of privacy is so ingrained in us. Because I am an optimist and believe that we can mature toward a more utopian world I believe we will choose to become more transparent and reap the rewards of this dynamic.

Humalogy Viewpoint

Estimating the Humalogical Balance between privacy and transparency is an interesting endeavor. Ever since the time that telephone operators could listen in on conversations and we invented tools like listening devices, we opened the doors to technology invading privacy. With the advent of the Internet and mobile devices grew the potential for a third party to know what we do. We also have those tools to thank for a movement toward self-selected transparency where millions of users supply information about themselves willingly to the world. Law enforcement has moved from listening in on verbal conversations to monitoring email and putting up video cameras to track what we write and where we go. This means that technology is providing the tools for us to have far less privacy and, at the same time, some people will use technology to try and defeat the systems that track us. For our purposes I am going to balance the use of human activities to lower privacy levels vs. technology.

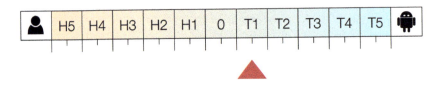

Although we have lots of new technology tools that impact our transparency and privacy we also still involve the human component quite a bit. Yes, the government has highly automated and sophisticated ways to monitor their technology based listening posts. but on the whole millions of people still invest their time either into posting information about their lives or who track information on others. In the future we will use more automation to "watch" what people are doing and alert us when we need to take action. We will develop highly sophisticated surveillance systems that can be used by a normal user to watch over our kids, home, pets, and other things that are important to us. For now I look at this area of our lives as having a T1 Humalogical Balance.

Seminal Questions

How deeply will technology go with gathering information about our lives and providing it to external parties?

Who should have the right to decide what should be private vs. public?

Will the world ever come to a consensus about what is private and what should be transparent, or will we have very different laws in different jurisdictions?

How much privacy should a person really expect to have in a highly technology-augmented world?

Would a much higher level of transparency help us be a more civilized and enlightened race? And if so is this part of God's plan for us to become more enlightened?

CHAPTER EIGHT:
Productivity, Automation, and Dependency

What happens if the human race invents its way right out of a job?

(A question asked of me regularly by audiences)

Another key impact of technology in the world is the ability to automate tasks that were formerly done, or that have never been able to be done, by humans. The 20th century was the turning point in this regard because we escalated our abilities to replace tasks done by humans with computers, software, robotics, and the automated workflows supported by all three at a dramatic rate. This expanded our ability dramatically to be highly productive, and from the 1920s forward we have seen a consistent upward curve forming with respect to our productivity rates. This has created a wonderful increase in the quality of life for people as our overall pay levels have grown around the world. There is now a dynamic called the Great Decoupling. This is the historical anomaly where we have had side-by-side growth in the levels of productivity and household income from the Depression era until approximately 1980, then productivity started going up at faster rates while income levels flattened out. Translated, that means we are working harder and getting more done but not being paid commensurate to the productivity. Obviously someone has to be enjoying the productivity rates, namely the owners of the organizations that are generating profits from our increased productivity.

In 1964 the most valuable company in the U.S. was AT&T. They were valued at $267 billion in today's dollars and employed 758,611 people. A good comparison of size and services is Google. The difference is that they are worth $370 billion and only have 55,000 employees. That is less than ten percent of the workforce AT&T employed. When looked at on a value-per-employee basis it is staggering the difference that technology-based automation and delivery makes on the employment of people.

This is important to understand because there are a lot of people who wonder what could happen as we automate the majority of tasks that humans do and create a situation where there are not enough jobs for people to fill. Would we then get into a situation where we further destroy the middle class and end up with a larger wealthy class who builds and owns all the technology? Correspondingly, would this create a world where the non-technology elite are reduced to doing whatever service jobs are left that have not been mechanized? Before we get into that discussion let's step back and examine our insatiable desire to find ways to use technology to do things that human beings can do, even if inefficiently.

Over the last two decades a huge amount of our collective energy and creativity in the economic world has gone into developing, funding, and implementing new technologies. There certainly is a financial motive for companies to create new uses for technology because one of two critical things results; either automation lowers costs or revenue is generated from people buying these new tools. Either of these positive impacts on profits dictates the direction toward technology for sure. On the revenue generation side, the upside potential of technology is proven by the fact that Apple rocketed from a niche player in the PC space to being the largest company in the U.S. based on market value. In fact, as of this writing, a glance at the fifty largest companies in the world shows that fifteen are direct suppliers of technology, with the other thirty-five being dependent on technology to produce their products or deliver their services.

As we yearn to improve our lives and our careers by applying more technology, we are moving the needle slowly on the Humalogical balance from the H side to the T. We have been in the process of this for centuries so the movement is not anything new. Although we think of technology in a computer and Internet sense today we had mechanical devices long ago and those technologies changed the game in their time. The simple act of planting crops, which was done by hand for hundreds of years, has advanced us many steps forward from the hand hoe to the computer-driven tractors we have today. So the process of growing food has moved from an H5 to what could be seen as a T3 today.

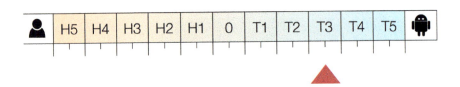

The difference now is the speed at which the balance is shifting toward the T side. If we had an ability to add together every process performed in this world we would certainly see that we are speeding up this shift. As our hunger for productivity speeds up further changes in the balance, there will be plenty of interesting changes for all of us; the rest of this chapter is focused on bringing those to light.

The first critical dynamic to consider is the impact on how we deal with time. Time has become one of the most important currencies we have. Over the past couple of centuries there has been a conversion from money being the resource we most sought to time being our most precious commodity. In the U.S. younger generations are starting to rebound from watching their parents and grandparents spend much of their time working. These young generations want to find a better balance and often are willing to make less money if it means they have more time at their disposal. Of course they have the luxury of doing this because the standard of living has become high enough that even a middle-of-the-road job provides enough income to live reasonably well. In economies where this is not true people are still willing to trade lots of their available time for money.

A recent Millennial Branding report found 45% of Millennials will choose workplace flexibility over pay.

Source: www.forbes.com

There is no question that now we are saving (or creating) huge amounts of time that is freed up when compared to old methods of accomplishing tasks. What is in question is why it does not feel like we have more time because most people will tell you they have significantly less time than in years past. How can this possibly be? How can we be so much more productive but be even busier than we ever have been? The answer lies in how we have chosen to reinvest the time that we have freed up. The reason the time dynamic is so critical to understand is that it is central to our quality of life. We get to choose how we use our digitally freed-up time, and what we choose to invest it in makes all the difference in the world. For better or worse the answer for many of us is that we have chosen to reinvest our found time in more work. (Sadly, it is Saturday night at 8:30 PM while I do the final edit on this book so you can see how I reinvest my found time!)

But if technology really does save us time it seems that the day should come when we get enough technology in place that we really will have more free time and have all the productivity we crave. I am sure we will get there, and the driver of this will be the further automation of tasks that are done now by humans. We have been on a steady march since the mainframe days toward using technology to automate formerly paper- and human-based processes. The pace of sliding from H to T on the Humalogy scale is picking up, and this is a wonderful thing. At the same time it scares a lot of people because of the possible negative impacts automation can bring.

Technology-driven automation drives critical changes in our lives and the economy. Because automation changes the very foundation of how work gets done it impacts the "who does the work" part of the equation. While we love when a process gets automated that helps us save time in our personal lives (like automated bill pay that the banks offer), we detest automation if it costs us our job. Because organizations have different motivations than the people who work for them, entities love automation because it brings financial benefits that cannot be ignored. Automation speeds up production, can lower errors, and can reduce costs. And, it also is less expensive than paying human beings to do a task over and over again. There will be people who worry about automation because of the personal impacts, and be assured we will automate everything we possibly can because organizations are not worried about an individual's job; they are worried about corporate survival.

So how do we answer the people who say, "What happens to the economy when all the people lose their jobs to automation?" I have heard this question asked by audiences many times, and it is a fair question to raise. If we automated the world to a high degree and there were not enough jobs for people, the global economy surely would suffer. So the basic issue to address is whether automation will lower the total number of jobs needed in the world or if the work that humans do will shift to other tasks. There is little research that we can refer to in helping us with what ultimately might happen in fifty years or more, however we can look at what has happened over the past twenty years and extrapolate forward.

Already we have automated many things since the early days of robotics and the Internet and the overall amount of jobs in the world has continued to climb. The standard of living across the world has risen. Yes, we have a shrinking middle class in the U.S. and yes, the rich are getting richer. At the same time the general quality of life for people in the world has risen even in the face of a high degree of automation. The Internet has helped raise the standard of living in many poor countries because we have been able to provide Internet access and mobile-based services even in underdeveloped nations. This allows people everywhere to take advantage of online software that can automate daily activities like banking, purchases, and education.

If we think of automation as augmenting what we do and not replacing us we can take a more positive perspective on the future. I believe that there will be a historically significant shift in the work people do versus what machines do over the next two decades. In the end I do not see the economy collapsing because there are not enough jobs. I see a future where the kinds of jobs people do are very different and we will come to understand that technology simply augment our ability to achieve and build so the kinds of jobs shift.

Let's go back to looking at the possibility that we will be able to go over the top of the hump and automate to the degree that we can work less and still be more productive. I have given this a lot of thought and want to share the following graph I developed in order to show you where I think we are along the automation/productivity curve with technology:

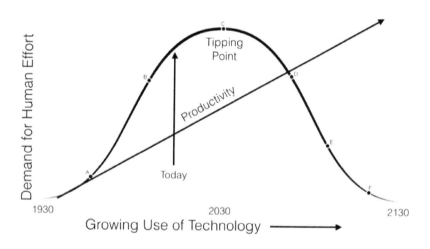

I believe that we are on a curve and that one day we will hit a tipping point where our technology improvements will catch up to our appetite for productivity. While we will seek constantly to be more productive we will hit a point where the technology will be providing enough of the work on our behalf that we will be able to work less while getting more done. Of course we have that situation now, but we keep reinvesting our free time into work in order to benefit financially. There will come a tipping point, however, where we will be getting paid well enough to fund the standard of living we choose and technology will have automated at such a deep level that we will feel comfortable working less.

Extrapolating further, there are a number of leading thinkers who are suggesting that, hundreds of years in the future, we will have a small minority of the population that actually works. In the past there was an ownership class that owned slaves and serfs who did most of the work. That left the ownership class to enjoy themselves or to pursue science and the arts as a kind of hobby or alternate (but hardly mandatory) profession. In the future technology will become our slave so to speak, and if that happens we will be freed from the daily drudgery that work can be today for many of us. We can own systems or robots instead that perform tasks and rent those systems out to organizations that need our specific technology capabilities. This is a crude analogy, I know, but the point is to be able to envision what might happen in a world where machines, devices, and software automate many of the tasks in our economy performed at the moment by people.

I think most of us take for granted the many small benefits of digital automation in our day. All of these seemingly small improvements add up to huge productivity gains in our lives because they free up minutes, which add up to hours. Also we become dependent on these daily improvements because the more technology does these tasks for us, the less we even have the skills to do them well. Let's take a look at some of these small things we do in our days that now have been impacted by technology and automation. To do this I have compared the past (and for our purposes think 150 years ago) with the present. Then I added the changes in the future I am predicting over the next thirty years or so. As you read through this list think about how the minutes saved would add up to hours.

Waking up in the morning: In the past we woke up with the sun by a rooster crowing or before the sun rose because work had to be done on the land or with the animals. We went to bed earlier because it was dark outside and light was not always easy to provide in a home. Because we went to bed earlier we got up more naturally and actually got more sleep.

Today many of us wake up with an alarm of some sort and possibly our mobile devices. We set the alarm based on a time schedule that often is driven by what we do for a living. We go to bed later because we have lots of distractions and the electricity available to make night seem like day as long as we want. Our waking time is not a natural time to wake up in most cases. It is not when our body tells us it has had enough sleep. We get up when we must get up to meet our first commitment of the day.

In the future waking up will be an intelligent decision that our technology makes based on a number of factors it determines. Our waking-up application will consider our calendar, map the drive to our first appointment or work location and then factor in our health goals that we have preset in our devices. Then it will factor in our past history for how long it takes us to get ready. After considering all of these factors and more it will decide when we should wake up. And how it wakes us will be more natural than just a rude sound-based alarm because our devices and homes will be able to coordinate efforts to open blinds slowly, begin to raise the volume of music, create vibration in our beds and so on. We will custom-develop our own waking protocols and the world will be a better place for each of us.

Preparing food: In the past we cooked over open fires that used wood we had to gather and chop as fuel. We used big metal pots that had to be scrubbed clean every day. We also had to grow or raise most of our own raw materials for the meals. The sheer amount of labor involved to produce the basic elements of a meal were extraordinary compared with today. In many cases a person in the family had to be fully dedicated to this task on behalf of the rest of the family. The food that we prepared might not have tasted quite as good, however it made up for this by being much healthier than the processed food we eat today.

Today few of us grow anything in a garden; we buy our raw materials in stores or eat out at restaurants. We still have traditional stoves and ovens but today they are not powered by wood we harvest by hand. They are electric or gas-powered. Further, we invented microwave ovens so, with the touch of a button, we can cook things very quickly. Because we are under time pressure with all our activities, we rely on "nuking" our food instead of cooking in healthier ways.

The latest craze is to have groceries delivered to our homes so we don't have to walk through a store to purchase them. It's likely we will create hybrid ovens that can cook with any one of four or five methods and even offer an option to combine any set of options to achieve either our time or creative cooking objectives. It is also likely that we will continue to eat more meals that are prepared by others instead of spending any time cooking at all. We want our food prepared faster, delivered faster and we consume it faster. Taking the time to prepare a family meal or just a home-cooked meal sadly is slipping away from too many of us.

215

The household cleaning: In the past we used brooms, rags brushes and lots of physical labor to clean the house. Our soaps and disinfectants were harsh and environmentally unfriendly. There was much time involved and, again, often there was a person who dedicated much of their life to this task on behalf of the family. Add in the job of cleaning clothes by hand in big metal tubs and a large family of people getting clothes dirty, and the household chores were quite an ongoing project. The description "back-breaking work" comes to mind.

Today we have Roomba Robot sweepers that can be left running when we are away from home. We have self-cleaning toilets that lower the load of scrubbing with a brush. We use more environmentally friendly cleaning products. We have invented high-tech mops to clean and disinfect at the same time. There is still a decent amount of human work involved in the house-cleaning process. Cleaning clothes has gotten much easier because we have high-tech washing and drying machines that cut the time and labor down dramatically, provided we don't just outsource all of this to a dry cleaner.

In the future the cleaning robots will take over more than just sweeping floors. We will use smart devices and robotics to clean our homes and offices and they will have enough intelligence to know what does not belong and to eradicate it. We will replace cleaning chemicals with nano-bots that will "eat" the dirt and bacteria, then be washed away. Cleaning clothes will get easier as we improve on our current nano-fiber based fabrics that kill bacteria, never get wrinkled and resist any kind of stain.

Transportation: In the past we walked or were transported by an animal. This had great health benefits because we were forced to use our muscles and build stamina. It also limited us greatly because we couldn't travel very far in most cases. We were out in the fresh air which could have a positive or negative impact depending on the weather.

Today we have automobiles, planes, and trains, and in each case we are improving how quickly and intelligently these machines can move people. We have unchained ourselves from a specific region and freed ourselves to work anywhere within sixty miles of our home by using our various means of transportation. In the larger picture there are many people who work virtually and not from anywhere near the corporate offices. This took away the desire for most people to use some form of exercise to move around and our health has suffered because of this.

In the future our travel systems will get faster and more self-guided. Cars will drive themselves to whatever destination we choose. Trains will give way to high-speed rail-based systems of various kinds. Planes will fly faster and higher so that long-range travel will get simpler and more practical. In all cases we will be freed from having to focus on anything while we travel other than work or each other. Consider for a minute how much more productive we could be if, when in a vehicle, we could work on whatever we wanted instead of paying attention to the road. We would reclaim billions of collective hours "wasted" on driving.

Finding information: In the past finding information or getting answers to our questions was extremely difficult. There were only two viable methods: asking a person who was physically close to us or looking it up in a book. The book option was highly dependent on our ability to get to a book that would have the information we needed. This meant if we were lucky we had an encyclopedia set at home. If not we had to go to a library. If we wanted a geography answer we had to find a map. If we wanted a medical answer we had to find a doctor. You get the idea. Not only was it difficult to get information; the chances that it was correct were pretty low depending on the source and normally we had only one or two sources.

As we exist today we have a rapidly expanding ability to get information anytime. We carry mobile devices that connect us to a huge source of information that is getting more refined all the time. When we ask for information we are connected to a vast array of sources so that we can invest the time to dig as deeply as we care to in order to verify accuracy. And there is information on virtually anything known in the world, already available on the Web. We have only had this resource for less than two decades and we are numb already to the fantastic nature of this.

In the future not only will accurate information be easier to find; it also will be simpler to access. The wearable devices we are inventing now will push information to us in multiple ways so that we can see it in a heads-up display or hear it directly in an audio device while doing work that occupies our eyes and hands. Further, information that we need will be pushed to us without asking for it. Software and our smart devices will watch over us and when they see that we are in need of help they will predict what we need and push it to us. Our issue will not be working hard to find information; it will be working hard to filter out the important things from the streams that are pushed to us.

Solving problems: In the past we had to be very self-reliant in order to solve problems. If something we owned broke, we had to be resourceful enough to find a way to repair it. If we had a physical problem oftentimes we had to figure out how to heal on our own. If we had a relationship or personal problem we might be able to ask a family member for help, but often we ended up trying to work it out ourselves or just lived with the problem the rest of our lives.

Today we have a staggering amount of help with solving problems. If something breaks around us we can go online and find the fix and, if needed, order the part for just about anything that was ever built. If we have a medical issue, we can go online and do lots of research self-diagnose and sometimes even come up with a remedy unless we need a doctor to perform a procedure. If we have personal issues, there are thousands of articles on every topic and online experts galore willing to help. The Internet has become the most powerful problem solving resource humankind ever has invented. The easier we can solve a problem, the less time we spend on this and the better our quality of life becomes.

In the future problem resolution will continue to get easier. In many cases the problems we run into will kick off automated workflows and initiate the corrections themselves. The devices around us will know if they stop working and will "phone home" to the manufacturer and begin the first steps of a possible repair on their own. In other words, many of our physical possessions will be self-healing or self-repairing. When we go online to solve problems that are not related to a broken physical thing, for example a billing problem with a credit card or a conflict with a doctor appointment, we will trigger automated workflows from their Websites that fix the problems instantly without a human being having to be involved. We will only have to take one easy step, and the rest will be handled on our behalf. If we do need a human involved in helping us we will use one of the many digital connection options that are more efficient than a phone call. All of these improvements will save us time and much frustration.

Communicating with each other: In the past we communicated by letter, word-of-mouth or carrier pigeon if we wanted to talk to people. We had a limited ability to talk to anyone who was not physically in our presence. If we wanted to talk to someone who was not in our vicinity we had to get on our horse and go see him or her, or we had to take the walk. If they lived many miles away we might never talk to them again. If we wanted to talk to hundreds of people at once the only option might be a town meeting and those were rare. If we wanted to talk to a million people the only option was to write something that was reprinted all over the country, and that was rarer still.

Today we have a plethora of options for communicating with people. We can "instant message" someone across the world if they have an Internet or cellular connection. We can videoconference with one person or many from any device we have that has a screen. We can communicate through text, audio, video, pictures, or any combination. We have multiple choices of communication platforms and the cost for all of this is dropping. Talking to any specific person anywhere in the world is possible now. We can talk to millions of people instantly and for free now as well.

In the future communicating will continue to be easier because the options that new devices will bring us will make it simpler to connect with a person instantly through any media types we choose. Communicating with each other is a high priority in our lives so the innovation in this field will be rapid. One of the most important things we must solve is notifying each other of our current availability to receive a communication and the best channel to reach us. We still have to guess about a person's availability today, so we reach out in whatever format we think is best depending on what we have to say. Soon we will be able to see a person's status in real time and will have better clarity about the best ways to connect. On the other side of this capability will be an improved ability to filter who can connect with us through all of our technology platforms. Because communicating with others normally is a big part of our day any efficiency we create in this area immediately pays off by saving time. So far, we have not used the time benefits given as we have actually just communication efficiencies to communicate more!

Entertainment: In the past entertainment was much simpler. A conversation on the porch was considered entertainment. Playing with sticks in the yard could yield a whole fantasy world of possibilities. Storytelling was an art form because stories were one of the most abundant and simple forms of entertaining each other. Before television families would gather around and read poetry or play music together which, as you can imagine, was a wonderful way to be creative and connect with others. These were ways to express ourselves through art and, without modern technology, we are doing less of this. Today that seems quaint to many people. Having neighbors or family over to the house for a meal was considered entertainment and if this included card games we were on the very edge of the entertainment spectrum.

Today we have elevated entertainment many levels past what we had even thirty years ago. With the advent of the Internet we gained a new delivery method for entertainment content and communication. We developed online gaming so we could "play" with people anywhere in the world collaboratively. We can download and play almost any song or movie created in history instantly and across multiple devices. We can play just about any game ever invented on our screens and can choose with whom we play them anywhere in the world. Now have the means to entertain ourselves at a moment's notice even for a few minutes at a time with our mobile devices. If we are given five minutes of empty time some people will fill it with a host of mobile games. A distinction to note is that too often we are consuming someone else's creations (music, art and stories) instead of creating them as more people did in the past.

In the future digital forms of entertainment will get more experientially realistic. We will continue to crave environments that feel real when we are playing which will drive virtuality to incredible levels. We will have full-immersion devices that will allow us to lose ourselves in our entertainment by suspending any sense that we are not at the simulated location. Part of the fun in being entertained can be losing a sense of reality, and soon this will get much easier. We will cease to be interested in playing Monopoly as a board game when we can play it in a virtual world where the game comes alive around us. We will get tired of having so many TV channels without being able to find a particular program whenever we want it so all video content will be on demand. We will be able to fill any amount of time with deep sensory entertainment and this will cause some to get quite addicted to these artificial worlds. This will not free time for us as much as it will give us new options for reinvesting the time we save from all of our automation.

In each case the changes listed above free up minutes and hours that we can reinvest in something else. Maybe some day when I want to write a book I will just dictate it to a computer that edits, formats, researches and organizes all my thoughts into a finished product. Maybe my computer will push out each page as I dictate it to the online crowd for their opinions and edits so the Universal mind has input as I create my book. My computer might also poll the Internet crowd as I am writing the book in order to test their interest level so I can see in real time what content is meaningful to the crowd as I write. If all of this would happen it certainly would not take me two years to write a book like this anymore.

Another dynamic to consider is the ability of automation to speed up processes far beyond what humans are able to do by hand. When we take people and paper out of the equation and replace them with technology we also naturally speed up the time to completion. This is great from a productivity standpoint and, at the same time, can be overwhelming for the people who have to collaborate on the processes that move much faster now. Although the brain is a miraculous organ and information-processing device, it also has upward limits on what each of us can absorb and deal with in a given span-time. By way of comparison think about the difference now and in the past when a person worked on a manufacturing line and put the same part on the same product over and over for hours and days at a time. That had to be excruciatingly mind-numbing in its repetitiveness. The equivalent activity today would be staring at a computer all day interfacing with information that is flowing by and having to make decisions at the speed of the system. While this may not seem as bad as a repetitive physical task, there are problems when a person's entire day is spent sitting in front of a screen using a keyboard and a mouse to manipulate data.

We have sped up the ability to solve business problems, sped up entire supply chains, sped up product development cycles, sped up taking marketing plans to the streets, and generally sped up our ability to communicate with anyone in our organizations. As we integrate ourselves with technology, we will aim to do the best we can to work at the speed of a digital network, and we will fail often. Ultimately we will exhaust ourselves trying to complete work, make decisions and absorb valuable information. Being able to complete any process faster and with quality is an amazing feat. However there is only so much we can handle before feeling overwhelmed.

There is a dynamic I will label the "self-tightening seatbelt" principle at work here. The seatbelts in vehicles are built to keep constant tension on our bodies; if there is any slack created the seatbelt ratchets tighter. Over time while driving, the seatbelt can become uncomfortable because it is always adjusting to the tightest fit. The technology in our lives is doing much the same in that every new device or capability makes digital functionality mandatory to us because we love the efficiency and convenience. Once we feel the results of the improvement we make them permanent because we have little desire to go back to our old way of doing things. Then many of us crave the next step forward because we appreciate the progress. We are developing an insatiable hunger for more functionality in the next version of the smartphone, the next device, or next version of software. The result of this spiraling up of automation and functionality in our digital tools is that we become addicted to a higher level of productivity. It speeds up our lives and it immerses our brains both with input and decisions to make. Sometimes I wonder if, like the seatbelts in our vehicles, we will get tired of the constriction and loosen the tension from time to time. Will we look to have times in our day or week when we want to step away from all technology just to be relaxed and free for a while?

Dependence on automating technology is not a problem per se, however there are two potential dangers we do face. The first is the danger of losing our ability to do a task "by hand" because we have depended on technology to provide it for so long. We may even lose the ability to understand how something was done originally, and at that point we don't even know if the technology is doing it right. Think about the GPS driving instructions on our cars or phones that we use now and how dependent many drivers are becoming on them. As long as we can read a map we have an ability to understand how to double-check or correct the application if it is wrong or out of date. When we lose an ability to read a map because we have become dependent on our digital directions we are stuck out in the cold if the app does not get us where we want to go, perhaps literally.

The second is the risk of huge productivity loss if we have a massive disruption of service and need to go back to doing tasks by hand. Hopefully we never face this situation but if it happens one day humanity would be crippled for a long time as we would be forced to relearn tasks we had dedicated to machines. There are plausible scenarios where a whole country or part of the world could have its technology wiped out overnight. One could be a rogue computer virus that has a destructive payload that gets into the world networks and cannot be stopped until it is too late. Another could be a war where one of the sides uses a weapon that would destroy electronics but not people. There is such a bomb called an Electromagnetic Pulse Bomb, or EMP bomb. The combination of a successful cyber warfare attack coupled with an EMP bombing could wipe out a vast swath of a country's ability to access the Internet. Let's not be naïve; the militaries around the world are quite aware that in order to harm an enemy the greatest damage that can be done might not be destroying property and people in a direct attack, but rather wiping out all their technology, rendering them helpless in many ways. This will especially be true after a few decades of relying on technology to do most things for us.

I hope we are able to avoid these levels of disaster and never have to take this kind of step backward on a great scale. We have enough to worry about with the impact of our reliance on technology and the productivity when it is working. The thought of the issues it would cause if it all came down are staggering. I have a feeling we will get to see this firsthand at some point because, sooner or later, one of the two examples above will happen and we will have our first case study on how people adjust to life without any technology. But for now let's get back to focusing on the impact of automation when it is working.

There is no better example of our love/hate relationship with digital productivity than the disappearing line between work and not-work. Now that we can take pieces of our job with us on any device we have the tools that allow us to work nonstop if we choose. That seemed freeing at first because it meant we could make up some work from home or continue to get paid even when we stayed at home with a sick child. It is liberating to be able to be productive sitting anywhere in the world and at any moment of the day or night. That liberation has turned a bit ugly for many of us, however. It has caused the number of hours we work to climb and, along with it, the amount of brain space that we are forced to dedicate to our career.

On the positive side, technology is allowing us to stand higher on the shoulders of those who went before us. The Internet, databases and software applications are combining to provide platforms that contain knowledge gained over the course of humanity. This amplifies the powerful dynamic of allowing each succeeding generation to start where the last left off instead of having to learn everything from scratch. This ability to stand on the shoulders of others is additive because it provides an easier path to prosperity. We do not have to start from scratch learning every lesson because we can search online and have the answer vetted by someone else delivered to us. It also makes knowledge cumulative because we can layer on new innovations to those designed by earlier generations and, thereby, make huge strides forward in the world. An example of this dynamic is scientists being able to improve on the work of those who went before them on a constant basis. We do not have to reinvent a solution to cure an infection because we already have antibiotics; we just need to improve on the concept of designing specific antibiotics for distinct purposes.

When I explain this concept of being able to stand on the shoulders of others by using technology to teach us or helping us through the programming and information stored by those before us, some people grouse that, then, we will lose our knowledge of how the underlying concepts work. They will point to the importance of understanding how basic math works and how it can be done by hand so that a person truly sees the whole picture. But does everyone really need to know how the underlying mechanics of something work?

We don't know how an engine works yet we drive a car. Most people don't know at a biological level why aspirin cures a headache, yet they are confident it works. Very few people know all the technology it takes to make a Google search work in a fraction of a second, yet they know how to use the search function. If we had to go back and learn all the underlying functions we could not invest in higher-level learning. So I believe firmly that for most people not understanding how something is accomplished at a basic level will be fine so that they can stand on the shoulders of those who automated or built systems before them. This multiplier effect is what will drive an escalating amount of progress.

Because software technologies store knowledge, data, processes, and automated functionality they will continue to grow in their capacities. Because devices will get smarter, cheaper, faster and more connected they will merge with software to provide fantastic examples of automation and, thereby, productivity. As companies provide their services and products for decades to come they will continue to build on the shoulders of the workers who went before and the pace of change, and innovation will spiral up. This simply was not the case in earlier eras when there was not a method for storing knowledge or automating the ability to do tasks in a machine. We passed knowledge, data, and processes down from human to human and often had to learn how to do them over and over through the generations. This created a very low rate of being able to stand on the shoulders of the previous generation and, therefore, a much lower rate of progress over the centuries.

What we will always have and always value are products that are made by hand and are unique and well crafted. There will always be a market for products not made by machines. The products we love the most in our lives might be products that were made by loving hands and with natural materials. These will balance out the mass-produced products in our lives. We will value the meal more highly that was made from plants grown in a local garden or farm. We will be in love with the coat that was made with wool grown on a sheep in the mountains that we visited and that was sewn by hand by a woman we met at the store. A large part of the economy will always be what is made "by hand" with natural materials by artisans. This will help provide the balance in our lives from the highly automated products and services we use for convenience.

I can illustrate the difference between something mass-produced from something crafted with this story. My wife and I both have Irish claddaugh rings as wedding rings. The one I originally got her for our wedding was yellow gold with a ruby and over time she asked

me if I would get her a different one with white gold and a diamond. An upgrade such as it was. I was glad to do this but I could not get comfortable with just buying one for her from a store or online. We looked a few times and she saw things that she liked, but it just did not seem right to me to get a ring for her that was mass-produced. I tracked down a guy who made rings from scratch to any design that a person could dream up. I had in mind a variation of a claddaugh ring with a heart shaped diamond and a diamond-encrusted band. We drew it up and then sorted through some diamonds until we found one of the right size and clarity. Then he had a unique mold made and poured the gold into it to form the platform for the diamonds. In the end, I was able to give her a one-of-a-kind ring that I had designed and a custom jeweler was kind enough to help me construct. Annette wears that ring proudly today and knows that it was designed, built, and delivered to her by a husband who cared enough to craft a unique symbol of love for her. This gives her ring a different level of meaning because it was not made by a stranger in a ring factory of sorts. It was made by hand by someone who cared about whether she specifically would love it.

I see a big split coming in the market between entities that are highly automated and productive and those that are not at all concerned about mass-production or efficiency because they hand craft unique products or services. To put a name on both, lets call them "Craft Economy" and "Efficiency Economy." We already see this divergence happening when we look at the difference in the business strategies between companies like Tom's Shoes and Nike. They both make shoes and have very different business models and goals for what they accomplish in the world. Think about the different business strategies used by Amazon and Etsy. They are both well known Websites, yet their approach differs in that Etsy is all about selling unique and often hand-crafted items, whereas Amazon is selling commodity items more often. In healthcare there are doctors who churn through patients at fifteen minutes a visit, but at the same time there are practitioners who are signing up patients on a subscription model so they can spend as much time as they would like with each patient.

There also will be a third type of economy which will be the "Maintenance Economy." This will contain all the work that is done to provide upkeep on property. It is likely that this will end up being the lowest paid of the three economies. For our purposes I will stick with commenting on the Craft and Efficiency Economies for the rest of the chapter.

Understanding the difference between the Craft Economy and the Efficiency Economy is important because this is what will allow us to automate many of the things we do in the world while still having careers for everyone who wants to work and be productive. As we automate the making of commodity products and services on one hand, there will be a balancing market for unique and unusual hand-crafted goods and services. We will love the efficient restaurants that provide food in seconds with no humans involved when we are in a hurry. We will love the craft restaurants where people get a completely different experience that is meaningful for them. We will love the inexpensive birdhouse that was mass-produced because it was cheap. We will crave the handmade wooden birdhouse that is unique as well because it is one-of-a-kind and we met the person who built it. This balance of the Craft Economy and Efficiency Economy will allow us on a grand scale to find the Humalogical balance that fits for each one of us.

Looking Forward into a Highly-Automated and Productive World

We truly have an amazing ability to invent solutions to improve just about anything we can possibly imagine. If we can visualize the perfect state of any situation we have the first step toward making it come true and, over time, we will iterate improvements until we get there. This basic human instinct to improve how we do things drives us to create and develop new ways to be highly automated and productive. Our economy is based heavily on providing us what we want because what we believe we have to have, we will pay for. This is different than 200 years ago when the economy was based more on what we absolutely needed to survive physically. Today we are focusing more on the needs of our souls and more on who we are being and not just what we are doing.

With that thought in mind let's look at what people will want going forward because what they want will drive how technology gets applied in the economy.

We want convenience: We like anything that will make our lives easier and, as such, anything technology can do to save time or work will inspire us to implement new digital tools.

We want lower prices: We want the cash that we have to go further so whenever a company can lower prices because they are able to run more efficiently, we reward their progress by purchasing their products. This drives a use of technology to lower operating and selling costs in order to pass savings on to us.

We want to work less: I said earlier that we are working harder and longer because of technology and that would lead one to believe that we do not want to work less. This does not mean that, given a choice to make a great living and work less, that we would not choose to back off a little. The moment we can get paid well and work less we will do that. We see the younger generations leaning this way already.

We want to achieve: Most of us find fulfillment in making progress, creating, inventing, and generally improving ourselves and our lives. We want to believe that each day has some value in it because we have achieved something new. Technology provides a host of tools that allow us to achieve. These tools can help us learn, create, and get things done faster. For this reason we will continue to strive to move forward by leveraging every new technology innovation we can get our hands on.

We want to stay healthy: Almost nothing else is going to matter to us if we do not have our health. We want to feel good and have a low amount of stress and pain. In most cases we also want to be as healthy as possible without a ton of work or discipline. New technologies like wearable devices help us monitor our health in real time. Social tools allow us to team with others over the Internet to inspire each other to achieve progress with our health goals. Search capabilities help us understand potential health problems and learn to make healthier choices.

We want to serve a greater purpose with our lives: As we are coming out of generations of struggling in survival mode, many more people are seeking to leave the world a better place. We see this in our new emphasis placed on companies that have a social mission. We see this in the millions of people who volunteer their time. Life is becoming less of a marathon of struggle just to stay alive for most people. It is becoming a chance to be a blessing to others and, in some cases, we are using technology to achieve this. When given a chance millions of people will use their mobile device either to donate small amounts of money or to pass on meaningful content that uplifts the world.

By understanding these six important human desires that drive our economy we can pretty easily forecast that productivity and automation will continue to be critical difference-makers for many organizations. Some will succeed simply because they learn to be as efficient as possible in delivering their products or services through managing a wonderful digital transformation.

Although we want all of the six areas that call for efficiency listed above, we also want a few other things and those are to feel special, to have unique experiences and to have meaning in our lives. These are the drivers that will help the craft economy to blossom. In many ways we will be stepping backward as we go forward because, two hundred years ago, almost all of the market wanted a craft economy and there is a lot to like about unique and hand-crafted goods and services.

Our drive to automate everything we can and to create as much efficiency as possible will continue to change how we live and how we work radically. Combine this with our deep-seated desires to achieve, to have more convenience and to work less, and we have powerful drivers that will continue to make technology innovations a very profitable business. The road forward toward a massively streamlined world will not be easy for everyone and there will be collateral damage with careers that get displaced by digital tools. Over time we will find our way back to an economic balance between the Craft Economy and the Efficiency Economy. Young people will have an improved ability to educate themselves with skills that will be valuable and not become irrelevant early in their careers.

The road to a highly-automated world will be rocky. There will jobs that will be displaced and entire industries will go away. Eventually we will find valuable things for all of us to do in the world. When we do get there we will have constructed a life where we will be able to work less and more efficiently, and that will be a very good thing for our quality of life. We will be able to use this automation to improve safety, the speed of information flow, the speed and quality of problem resolution, and our ability to design the tools around us to efficiently operate in the ways we choose as well. This sounds a lot like a better world to me.

SUMMARY:
Productivity, Automation, and Dependency

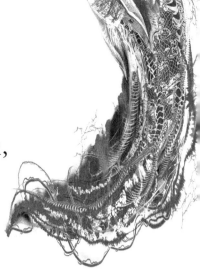

Our ability to automate many parts of our lives and the economy has far-reaching consequences. We are able to accomplish more in a shorter amount of time. We are able to continue the march away from paper as a storage medium and move toward databases. We are replacing humans as factory floor workers. We are also replacing many jobs that are based on making repetitive decisions that software can now make just as well or better. Now we are building companies that are worth billions of dollars with just a small team of employees (Instagram sold to Facebook for approximately $1 billion and had thirteen employees). Computers, robotics, and software will continue to replace tasks done by humans. There are many great reasons for this to happen and there will be large adjustments to be made in the economy.

I am sure we will ebb and flow as to how we feel about the collateral impacts of automating things that humans have done. How much heartache is involved in the transformation will depend on our specific personal experiences with how technology changes our life. Organizations will continue to push the envelope with automation because it is good for business. For a while longer all of this automation and efficiency will put more of a burden on workers to use technology to provide value for their employer. At some point we will hit the tipping point and will be able to work less and still make more money. People will be able to leverage their human gifts (creativity and relationship building) to add value that cannot be provided my machines and will be well paid for it.

My final call on technology and productivity is a 70% positive impact on humanity and a 30% negative influence and that is temporary in nature. We will suffer temporary setbacks when job types get displaced. We will struggle when we come to depend on technology to work for us and it breaks or goes offline. In the end, however, the world will be a much easier place to live in just as it is today when compared to hundreds of years ago. The use of machines to make life easier will not stop. Over time I see the automation of tasks as freeing us from doing things that are repetitive and non-creative. The world will be a much better place with technology doing things that are not uniquely human skills.

Humalogy Viewpoint

What is interesting to see on the Humalogical balance is where we are historically with machines doing things for us versus humans. It is interesting too to speculate on where the perfect balance should be. If we take all the tasks that are done in the world, how many of them can, and will, be done by technology of some sort and how many will be done by humans? What could happen to us if we move too far toward T and replace critical tasks humans need to do with technology (like if robots become our friends to fulfill that need)? At this time in history I believe we have moved from an H5 world when we first started developing technology to an H2 today.

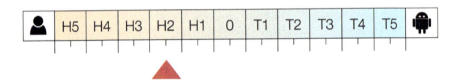

I am sure some of you might disagree and feel we are over on the T side already. I am factoring in everything we still do as humans today that someday will be done by machines and software. We still perform thousands of tasks that we will not need to do in the future. I believe that means we have a long way to go before we find the balance point that is utopian for humanity. And what might that score be? I suspect that the 0 score, denoting an equal balance, is the place we will want to be. I know that obscures the vision of a massive womb of automated technology that we would live in. When really factoring in all tasks done I still think there will be a huge Craft Economy where we value things built, grown, and provided by human hands. We will love the automation that helps us work less and have a higher quality of life. We will love the uniqueness of something done by hand too. A massage given by a machine will loosen our muscles in the same way a massage given by a person does. They will both have their place in our lives.

Seminal Questions

Will high levels of digital automation and productivity ultimately create a nirvana for us where computers and robots do most of the hard work humans once did?

What will we learn is the unique value of a human who is distinct and has a soul to guide them versus an extremely intelligent machine?

As machines and software take over more of the tasks we once did will we completely lose the ability to do them? Will this ever be a problem for us?

Will the Craft Economy and Efficiency Economy be of equal value in the world or will one dominate the other from a financial standpoint?

Ultimately will we delineate the world by tasks that are human-specific and machine-specific: face-to-face communications, building relationships, creativity, trust building for humans and just about everything else being done by robots and smart software?

Will this highly automated and efficient future help us have happier and more fulfilling lives?

CHAPTER NINE:
Mobility and the Outboard Brain

There was a time when we thought one brain was enough...

We took a giant step forward with technology when we moved it from mainframes to personal computers. We did it again when we connected all of our computers with the Internet. The next huge step was to take our connected devices and make them portable. That we still call them "mobile phones" is a misnomer because the percentage of time we use them as a phone is small as compared to their use as a computer. With these innovations we have introduced a powerful amount of computing capability into our daily lives that we use all day and night from just about anywhere we would care to go in the world.

The most interesting impact of having a mobile connection to the Internet is that our devices are turning into outboard brains. The first time I read an article written by a psychologist suggesting this concept, immediately I knew we would continue to strive for this outcome, even if we did not call it such. Set aside the role a smart phone plays in our lives today and just think about the capabilities we really crave. Given a choice we want the ability to ask a question instantly and have it answered accurately., I If that answer could be imprinted into our brain, even better. We want the capability to store huge amounts of information that are far beyond what we can memorize, and to have instant access to any of it when we need it. We want an on-demand ability to solve any problem that we face by accessing technology, helping us find the best solution, and even to help address it if possible. When all of these capabilities are built into a device, that technology truly is acting as an outboard brain.

We are falling in love with having an outboard brain because we are finding that it helps us be smarter, solve problems faster and better and it connects us with people around us in wonderful ways. The scary, wonderful thing is we have only had this capability for a short time. The fact that we call our mobile devices "smart phones" is no accident. They are smart and getting smarter all the time. They are starting to get to know us and adjust to our specific needs and wants. This includes being able to sense how we feel, the problems we often have, and kinds of information we need most often. They monitor where we are physically and can make suggestions about how to get to our destination. They watch our calendar and try to help us to stay on track. They even connect with other devices in our lives to help us have control over them. We can ask a smart phone questions verbally and they can answer us in the same way.

There are three critical elements to how a mobile device augments what our brain does for us. Let's go a bit deeper into each of these:

They memorize information for us: Studies are showing that young people have less of an ability to memorize small bits of data. This is likely due to the fact that they are not practicing this because they do not have to memorize phone numbers, birthdays, addresses, appointments and the like any more. All of these are stored in their mobile devices and, for the most part, never will be "forgotten" by the device. We are growing in our trust, storing information in the cloud and on our devices, not worrying about forcing our brain cells into the situation of having instant recall of this information. Our brains always have had a pretty good ability to memorize information, however it is simply not as exact, scalable, or dependable as silicon-based memory. It is also helpful to us to not have the pressure of memorizing what could be insignificant data when it can be stored in our outboard brain instead.

They help us solve problems: At increasing rates we are being given mobile applications that solve problems in our lives. These applications help us save time, find information, know what to do in an emergency, and be able to connect instantly with another person if we need help. In fact the driving factor for many parents to buy them for young children is to solve a number of safety and communication problems that can occur without the devices to assist. If two people disagree on a fact we can end the argument immediately with a search. If we have an emergency we can learn how to fix it ourselves or contact multiple parties. We are becoming so confident in our ability to solve problems with our mobile devices that we are anxious to be away from home without them in many cases. Our brain always has been important for solving problems; now we have a new tool that is proving very valuable.

They provide access to huge amounts of knowledge: In the past when we needed a piece of information either we had it in our brain, had to find someone close by to ask or found our way to the library. Today we have nearly instant access to a vast percentage of the human knowledge base. We can search by text or we can ask our devices verbally and get answers as if they were a very intelligent friend. The ability to tap into the Universal Mind to get information is a game-changing capability of a smart device today.

None of us made a conscious attempt to augment our brains with a mobile device; it just happened. This was not a conscious goal, but rather more of a wonderful happenstance. The floodgates are open now because we have learned how valuable it can be to have a powerful tool on our person at all times. Our outboard brains provide the capability to extend our natural brains' capacities and in many ways that are ancillary to the three already mentioned. Because we can search and consume massive amounts of information in the nooks and crannies of our days we can learn more about the world at a faster rate. Because they provide new capabilities for entertaining us we can kill time in any number of ways with games, videos and music too.

Mobile devices gave us multiple tools for communicating with other people individually or in groups. Not only did it unchain us from the wall; it also expanded our inventory of communication tools so that we could have verbal or text-based conversations in many different forms for specific reasons. We can text message a single person with a single question or thought, or we can tweet the world with the same.

The fact that our mobile/wearable devices are becoming outboard brains is not the only reason they are so important to us. The cornucopia of new services coming available, in a very real way, change our lives substantially. Let's take a look at a partial list of some other capabilities that have been provided to us with these new tools:

Instant entertainment: If you have children you have probably seen that, even with two minutes of nothing to occupy their minds, the mobile device comes out. Take a look at people who are standing in line and note how many of them will be on their devices. Whether they are "playing" on their device or just looking at Facebook or Twitter, they are using the device as entertainment. Some people spend hours playing games that have come and gone like Bejeweled, Candy Crush and Angry Birds. In a world where boredom is now optional, a mobile device filled with entertaining applications is a very useful antidote.

Instant transactions: Feel the urge to buy something? Now you are just seconds away. I have watched my wife do our Christmas shopping from a mobile device while sitting on our couch. More than that, I have watched people sitting in traffic (not driving of course) search and find an item that they had just seen and make the purchase without having to put their credit card in the system because it is already on file. Not only can we do mobile e-commerce at a moment's notice; we can also use our mobile devices as digital wallets so that we can make doing a transaction in person easier and more secure. There was q time not long ago when, if you wanted to buy an item, you had to go to a store, find it on the shelves, stand in line and the store had to be open when you wanted something. Today you can buy just about anything you want from anywhere in the world and it will be delivered as quickly as you feel like having it shipped.

Connections to organizations: Once companies were given the ability to build custom mobile applications a completely new relationship could be formed between brands and their constituents. Companies like Starbucks, Great Clips, Home Depot and American Airlines have millions of customers who have downloaded their applications so that services and information can be readily delivered. The applications themselves are more than just a collection of features; they are an extension of the brand itself. These applications create brand loyalty and, in some ways, humanize the company to a greater degree. The organization is not just a building, office, or store now; it is a "friend" on our phone providing help or access to the company in new and convenient ways. As the user gets more comfortable with the brands application on his or her mobile device, there is less of a chance they will load a competitor's tools. In fact, now there are situations where the customer relationship with the organization is all through the mobile application and not involving a location or a human at all. Great examples of this are the insurance applications that allow a customer to buy the insurance, enroll, file a claim, and contact customers service all on a mobile self-serve basis now without talking to an employee.

Geographic awareness: Most mobile devices have a GPS built into them so they know exactly where they are in the world. By sharing this information with applications, a user can easily "see" what is around them if need be. GPS also allows us to use our devices as real-time guided maps, to find a restaurant nearby or the closest hospital in an emergency. This is a very powerful capability because we have only had street signs and maps to help us in the past, and those were not always useful in a strange place. Not only do we know exactly where we are today we also can know where our family and friends are in relation to us. The restaurant business is being changed dramatically by the fact that someone can check and see where the closest Thai food place is and then make a choice based on the number of recommendations it has received. Our devices can tell us where they are when they get lost or stolen. They can also tell brands we trust when we are close to their locations (through geofencing) and facilitate ads and discounts being sent to us while in a store. There is a lot to like about the fact that our devices have geographic awareness.

Pictures and video: Our devices have an ever-increasing capability to take high quality photos and video. This has spawned a number of new dynamics that span the gamut from very useful to absurd. Combining built-in cameras with social applications gave us the concept of taking and posting selfies as a way of telling the world what we are doing at any specific moment. The positive aspects of having the ability to take pictures and video include being able to chronicle one's life in a much deeper way and to share events with others in real time. Pictures and video capture memories that, in the past, might have gone undocumented. On the negative side, these same technologies have enabled young people to send nude pictures of themselves to boyfriends or girlfriends who then pass them on to friends and possibly around the world. In 2011 Instagram analyzed what kinds of cameras the photos were coming from that were being posted on their site and for the first time, the iPhone was more popular as a camera than traditional cameras. Since that time mobile devices have been getting higher quality cameras and filters built in so now they dwarf traditional cameras in the volume of pictures taken. This powerful capability has also helped the next areas I will cover…

Citizen journalism: Mobile devices with cameras and connections to social sites have armed everyone who owns one with the ability to record an event or idea and share it with a billion people. Once uploaded the audience size is only dictated by how many connections the person has and how viral the content becomes. Because smart phones are so prevalent now it is a safe bet that there is always at least one person within shouting distance of just about anything that happens in the world. A large number of the photos - and now video - that are taken during an event are provided by citizen journalists who just happen to be quick on the button. This is having a major impact on society because this news is completely unfiltered, instantly available and, in some cases, quite raw. A recent example of the impact is all the video recording of police brutality or shootings that now are being captured by citizens. In the past these events happened across the U.S. and we normally only heard about something in our city. If it was just a story without a video to watch along with it, we might be unhappy but we were not too emotional about it all. Things are very different when we see video of police behavior now almost every week and we can make our own judgments about the propriety of these actions. Granted, the police have a difficult job. I also know that we are seeing too many instances of over-zealous use of force, and the fact that these are recorded is allowing us to have a great discussion on where the line needs to be drawn when enforcing the law.

Citizen journalists can also use mobile tools to deliver opinions the instant they have them. Instead of waiting for days or weeks to get an editorial comment or piece out to the world through traditional media these thoughts can be delivered to millions of people within moments of memorializing them with a mobile app. The impact of this cannot be understated because more popular citizen journalists can reach an audience as large as they might have with traditional means like magazines or newspapers, and they can do it without cost. This has loosened the control of traditional media on editorial opinions.

Add to this that citizen journalists can "publish" from anywhere in the world, to anywhere in the world, and we have a powerful new way to fight tyranny or bring the world closer together by sharing information from abroad.

Health and wellness: Another extremely valuable use for mobile/wearable devices is tracking our health and wellness. This includes the number of steps we take, the calories we eat and how well we sleep. For those of us who use them this way, the impact is clear and measurable because it is easy to see day-by-day and over time how we have improved. Just knowing that our device is measuring our activities drives positive change. I have developed a habit of making sure that I get at least 10,000 every day because of my health apps, which has made me very aware of many little tricks to get in the extra steps. The sleep monitoring capabilities have taught me a lot about how well I sleep and how different I feel depending on my ability to get the proper amount of deep sleep. The exciting part of all this is that we are just getting started. Every month a new device comes out that has more functionality than the ones before. It is a reasonable expectation that, within the next decade, we will have a very wide range of health analytics available to us. Once we have a deep stream of data coming off our bodies, there will be companies lined up to access this information. The early providers of services that use this data will be health and life insurers, doctors, hospitals, and health clubs.

I have only listed seven of the major categories of functionality our mobile devices provide today. Aside from being our outboard brains, mobile devices also have become our televisions, game players, safety nets, personal trainers, digital scrapbooks, maps, and digital wallets. When looked at through that lens it is no surprise why this device has become so important to many of us.

½ of all adults surveyed said they sleep with their phone.

Source: telenav.com

In order for us to have all of that capability we need the device to be connected to the Web. We need to be "on the grid." When on the grid we have a dazzling array of functionality that enhances our lives in incredible ways. There is a lot to like about being on the grid especially if we are working, in school or worried about our safety for any reason. Some of the functionality listed earlier can be done without an Internet connection but much of it is dependent on being connected. This brings up the topic of what it means to be off the grid. There are three reasons for being off the grid. The first is not having a data connection. The second is not having a device physically in our immediate possession. And the third is having a device that is broken. Only one of these is normally by choice.

I remember the first time I felt like I was on the grid. I was driving in my boss' car which I had borrowed to take a long highway drive to a client's site. On the way back I needed to make a call in order to secure an order for 20 IBM PCs. He had a bag phone in the car and told me I could use it if I needed to make any calls. I picked it up, cradled it on my shoulder and called my source for the PCs. Within a few minutes I had completed the transaction. I remember every detail of that call and how excited I was to pull off an order like that while driving down the road. It was my first real sense of mobility and the power of being productive while driving down the highway. A few days ago I drove to Dallas to give a speech. While on the long, straight highway I flipped through my presentation on the laptop, checked my email, downloaded some graphics, made a few calls, and basically got all the work done I would have done normally while in the office. Don't ask if I was driving because I refuse to answer that. Being on the grid has changed my ability to be productive dramatically, and I am addicted to progress. I am not addicted to my devices but I am addicted to getting things done.

Here is a story about being off the grid: I left home on a business trip of three days last year and realized while I was on the plane that I had left my iPhone at home. My wife offered to overnight it to me which

I appreciated, and I decided instead just to see what life would be like without it for three days. There was always this vague feeling that something was missing in my life. It was a tangible sense that I was missing something and, no matter how hard I tried, I could not make it go away. I enjoyed not being tethered to my device for eighteen hours a day because things were a bit more peaceful and, at the same time, I subliminally counted down the hours until I got it back. If I would have been on vacation being off the grid would not have bothered me. Working without my device did not feel that comfortable, however.

While I believe there will always be a fraction of people who, for various reasons, will cut out mobile devices from their (and their children's) lives, the vast majority of the world will bond with mobile technology in a deepening way. The decision will not be so much about liking or disliking mobile technology; it is more about being connected to the Universal mind and all the services available or living off the grid in order to gain the benefits of not being connected. And there are benefits both states depending on a persons needs at any particular time.

As difficult as this may sound for some people reading this book, having the confidence in life to put your mobile device down and walk away is not a bad thing. This is a debate I get into often with people because they always look at me perplexed and say, "Why does it make any sense to be away from my smart phone when the whole reason I carry it is to be available if my kids, my parents, my friends or others need me?" Sometimes they will argue further, "What if I have an emergency and need to contact someone immediately or find some information quickly?" I can truly see that they can find no logical reason to put the device away for a time.

Although I believe these arguments are reasonable, the constant attachment to a device does have its negatives. When we are on the grid we have no ability to control interruptions. There are times in life when it is healthy to be able to focus on a task without the threat of a Pavlovian-trained sound alerting us to bad weather, a tweet, a Facebook post, a text message, an email, a sports update or a phone call. I have been an avid observer of this in my own life and am amazed at how difficult it can be for people around me to walk away from this barrage of alerts or to ignore them when they happen. How can a person meditate, slip into deep contemplative thinking, have a meaningful conversation with another person face-to-face or even get some sleep with constant digital interruptions?

Speaking of interruptions, there is a big distraction factor with being on the grid with any device – laptops and wearables included. Regardless of the type of mobile or wearable one has on, the constant alerts and the desire to respond to them takes our minds off of anything else we were doing. We are having lots of problems with people driving and interacting with their devices and I am sure this will only get worse as we have more wearables to add to the mobility we have now. Driving is not the only task that we do that is compromised by the pull of being on the grid. Sometimes we cannot even muddle through a face-to-face conversation without our mind being pulled in another direction. It can be hard to study and learn when we cannot focus our minds on the subject at hand without distractions. I see people who cannot make it through thirty minutes of a church message without spending part of it on the grid. And no, they are not always looking at the Bible online.

I have been wearing various types of smart watches for almost two years now. I have a love/hate relationship with them because the buzzing notifications on my wrist can be handy to let me see instantly that someone is trying to contact me, that I have an appointment or alerting me to some interesting news that has hit the Web. At the same time it is hard to focus on a conversation with a person when either I have to ignore the buzzing on my wrist or look down and be absorbed in the message of the moment instead of the task at hand. I suppose this is a situation that each of us will have to decide for ourselves how to handle as time goes on. I wonder what we will consider rude and what we will consider normal ten years from now. Social mores and norms shift over time. In the recent past many people felt like talking on a mobile phone in a crowd of strangers was rude and it is bow becoming much less so. I am sure that some of the actions young people do with mobile devices today will seem completely normal in a decade.

One dynamic of our mobile devices that nobody really saw coming is the dramatic change in the level of privacy users have. Mobile devices provide the capability for other people or organizations to know where we are, what we are doing at that moment, who we are with and even what we are thinking. Earlier, in the chapter on privacy, we talked about the three types of privacy: body, information, and activity. Mobile devices can have an impact on all three. Body privacy can be invaded by cameras built into phones. Information privacy can be invaded by third parties grabbing information from applications of which people are unaware. Activity privacy can be invaded if the information about where someone is and what they are doing is monitored on their phone. There is a fine line between the safety a mobile device provides by being on our person at all times and the potential to use it as a monitoring device without our permission. How we feel about mobility and privacy has a lot to do with our personal belief systems about privacy in general. Let's look at three different types of people as it relates to privacy and mobile tech:

The Privacy Fanatic: This person does not want anyone to know anything about them and they will barely use technology of any kind because they fear what can be tracked. They go out of their way to not provide information, to create aliases, and to not use any application on a mobile device that would give away their location. In some cases, they will only carry a feature phone because they do not generally have a GPS built in.

The Privacy-Conscious User: This person has no interest in the world knowing what they are doing. They will use the Internet and will carry a mobile device yet will not use social media that provides too much uncontrolled access to anything posted. They are fine reading other people's posts but have no intention of joining the fray. They will turn off an applications' ability to use their location even if it cripples the performance. The risk simply is not worth the reward to this person. They will fill out minimal information on a profile page and will do anything they can not to to provide an email address if possible. This person cares about information privacy, activity privacy, and body privacy.

The Privacy Minimalist: This person is a high-frequency poster of pictures, videos, thoughts and information. They are fine with the world knowing everything about their public life. They have no problem with applications using their location and, in fact, will post pictures and make comments about wherever they might be which paints a clear picture of where they are at the moment. They will fill out complete profiles online so that others will know whether they have anything in common. Their main need for privacy is body privacy.

The Fully Transparent User: There is a small but growing group of people who are interested in "life casting." This is the real-time broadcasting of nearly every detail of their lives online through video feeds. We have a growing list of tools that will provide a real-time video to the Web so we will also have a growing list of people who over-share the details of their lives. In the non-video world we have some bloggers who will write about the most personal details of their lives. The reason this small faction should be interesting to us is that they may be proving that full transparency does not destroy their lives. To the degree that more people become comfortable with full transparency not being dangerous, it may pull us dramatically away from the high levels of privacy the older generations were taught to value.

Mobile devices provide powerful tools to enable the Minimalist and the Transparent to connect with brands and people. Mobile devices perplex and annoy the Fanatic and the Conscious. What cannot be ignored is that using a mobile device at all lowers our privacy at some level because companies and governments have the ability to track what we do on your devices. They also can monitor where we go with them and it seems we have little say in this. If we truly want to avoid anyone knowing these kinds of things about us, mobility should be avoided altogether.

There is an interesting trade-off we make consciously or subconsciously about how much we are willing to share with others versus what we want to keep to ourselves. In order to connect with more people more often, we are willing to trade off privacy for transparency. In general people seem to be moving toward more openness at a pretty rapid rate. Because younger generations are highly populated with Privacy Minimalists, it is a safe bet that the concept of what needs to be private is going to be radically different in the future.

A positive impact of sharing with others through our mobile devices is the power of the crowd dynamics. Many of the mobile apps we are using are powered by the "Internet herd" in some way. For this reason there is a huge benefit to becoming part of the collective, sharing our thoughts and leveraging the thoughts of others. Whether we are choosing a restaurant, solving an algebra problem or looking up something on Wikipedia, we are using information provided by the crowd. This capability will multiply as more people get proficient with mobile applications. As more people are rating, commenting and collaborating, apps will become more accurate, useful and powerful. On the whole this will be a wonderful thing for us as long as we are mature enough to handle what we are gaining by standing on the shoulders of others. This is a fantastic example of the value of oneness and how connecting across generations will help us grow to enlightenment more quickly.

What I tell people all the time is that a 14-year-old with a mobile device today is much more capable than a 14-year-old from fifty years ago. Using a simple search, a teenager today has an amazing ability to repair things, solve problems, find any piece of needed information, communicate instantly with anyone to provide help, and find something they need within a few miles. Decades ago a teenager might have a better command of hand tools, might be in better physical shape and might even be better conditioned to work harder. But compare them side by side and I would bet on the teenager today achieving a list of tasks more efficiently and effectively. At the same time take the mobile device away from many of our teenagers today and they are crippled in their abilities to achieve even basic things one could do a few decades ago.

This is only going to become more pronounced over time, and the question to answer is how we feel about that. Should we start teaching young people survival skills without a mobile device? Should we have classes in school that expressly forbid using technology as an aid? Will we send kids to mobile-less camps one day where they have to complete challenges without a mobile device? Although this sounds reasonable I seriously doubt it will happen. Here is my logic:

1. When we gained the ability to drive in cars and fly in planes we did not make our kids walk miles to school just to make sure they could do it. For safety reasons and because kids expect to be driven most parents would not let their kids walk three blocks to school.

2. When we got calculators we did not have "by-hand" math classes that forbid using them so students would have to do the entire underlying math. After junior high people use calculators the rest of their lives and most forget completely how to do math by hand.

3. We have had television and digital games for the past fifty years and have not told kids to turn them off and go outside and play, even though it would clearly be healthy for them. For the most part we just let them entertain themselves.

Once we have a tool that helps make our life easier and safer we become dependent on that tool and never look back. We understand that the tool might make us mentally weaker, less healthy, and less self-sufficient, but because we are "better" with the tool we weave it into our lives permanently and deal the best we can with the unanticipated impacts. What is different this time is that mobile devices are so powerful that weaving them into our lives has enhanced us in fantastic ways. At the same time they can weaken our abilities to be self-sufficient if we ever lose the use of them.

The concept that mobile devices will be our outboard brains is a game-changer for humanity. By putting computing power closer and closer to our physical selves we will further utilize the power of the Web and the Universal mind. One day there will be many books written on this subject and it will seem like our history without an outboard brain was like another Dark Age. I appreciate that I will get to live parts of my life with and without an augmenting brain because it allows me to have perspective on the advantages of both. One hundred years from now there will be very few people who will even consider living without technology augmentation. This concept is not only exhibiting powerful ramifications today; it will continue to gain traction.

The Future of Mobility

We have combined mobile devices with Internet connectivity and turned them into what we use as outboard brains. These devices augment the role our brains play in our lives either by enhancing our access to information or by offloading information we do not need to memorize anymore. We are finding this augmentation to be a wonderful addition to our lives because it makes us smarter, makes life more convenient and provides safety for us as well. The devices are able to augment our brains because they are connected in real time to the Web and to anyone else (or any group) we choose to be connected with. We started with a mobile "phone" and now will splinter into many different kinds of connected devices to give us even more powerful outboard brain capabilities.

We are on a path to mature from mobile to wearable and then to implantable technologies. Already we have all three in in our world, so this is not a difficult prediction to make; this is simply trend extrapolation. We have been building mobile phone/computing devices since the 90s and started selling consumer-level wearable computing devices around 2010. Presently we are deep into research and experimentation with many types of implantable devices that will give us a greater level of control over devices like prosthetics.

Here is a quick primer on what appears to be a fairly obvious progression for each:

Mobile devices: The hardware capabilities of handsets will continue to improve in a few basic areas. They will have faster processors, more memory and better screen configurations. In each case consumers crave each of these because they tie directly to the performance of the device. In addition they will change in shape and size, get thinner, lighter and more customizable. This will ensure that we see an endless stream of upgrades. What will be more interesting will be new capabilities for interfacing with our devices. We will continue to improve the ability to control them with our voice, hand gestures and with options for touch control. Soon we will see devices that are increasingly able to sense who is holding them, and this will be a great security feature. It will allow devices to completely reconfigure based on who is holding them so that the owner gets full functionality and so a thief gets nothing but caught. Because the device will be able to "know" who is holding it we will be able to set up guest rights for our friends if they are holding the phone.

Through their close connection to our voice patterns, skin temperature, and force of our touch, these devices will sense how we are feeling. Once the device determines our state of mind and our emotions related to that, a whole world of new functionality will open up. If we are angry and stressed our device will adjust how it operates in an attempt to make things easier. There will be a whole list of activities it will monitor to assure that we don't do anything rash. This could include asking us to double click to makes sure we want to send that flaming text message. Our outboard brain might also take preprogrammed steps to calm us down like popping up reminders to breathe or giving us a message that the world is not such a bad place. As time goes on we will have loads of options for configuring our devices to help when we are not in a good mental or emotional state. This will make them feel more like friends that pieces of hardware.

This is a good example of how our devices will become more self-learning. They will study how each of us operates and will customize the capabilities delivered. This molding and customizing process will become critical so that our devices are as efficient as possible in helping make our lives easier. The biggest improvements we will see with mobility not only be the hardware; they also will be in the software applications that provide most of the functionality.

There will be a constant stream of applications that help us automate tasks in our lives. There are hundreds of activities we do each and everyday, and there will be options to make each one of them easier. When we need food delivered it will be one click away. We will not have to look up menus or restaurants; we will just tell the phone what we want and it will be delivered and paid for automatically. Every industry will build and improve mobile tools to allow us to "self-serve" our own capabilities with as few clicks as possible or with a verbal command.

We will come to be judged not only on our human capabilities alone but on our ability to integrate with our mobile devices to produce results as well. When a person is applying for a job the interviewer will want to know the level of technology integration of the applicant. This will be a fair question because there will be a dramatic difference between someone who can use mobile and wearable expertly and someone who has no skills in this area. The variability in mobile usage capabilities will put huge pressure on schools because students who are well integrated with their devices will be able to learn at a much faster rate than those who do not have access to real-time information transfer.

As early adopters and technology pioneers prove that expert integration of mobility into their lives gives them an advantage in the world others will follow. The frightening aspect of this is that millions of people might march blindly toward being cyborg-like without understanding the impact until it is too late. Today this trend looks like a person staring at their mobile phone while walking along, completely unaware of anything else around them. Tomorrow this could look like a crowd of zombies with our brains engaged by an immersive stream of mobile information projected onto our retinas.

All of these device performance improvements will cause us to build more personal relationships with our devices because they will be further customized and play a more integral role on our lives. They will cease to be thought of as "phones" and we will stop using that term. They will be more like mobile assistants with avatars built in that are tuned to the way we need them to function. We will get more dependent on them and, at the same time, many will relish the time when we can disconnect.

Wearable devices: When we think of wearable technology today we often think of a device on our wrist or possibly a heads-up display of some sort. Soon we will see an explosion of various types of devices, clothing and jewelry that will be tuned to deliver specific functionality. There will be hearable devices that look like hearing aids and are connected to the Internet. They will talk to us in order to provide needed information audibly based on where we are standing or what we are doing at the moment. There will be many different types of heads-up displays, goggles, Virtual Reality glasses, retinal displays, haptic gloves and others. In all cases these will be task-specific because different uses will call for different types of capabilities. The wearable we use at work will be different from the one we use to play games. There will be specific units to help surgeons and different models to help computer technicians. In all cases the heads-up displays will push information into our eyesight to augment what we are looking at in the real world or to replace the real world with a virtual world.

There will be a whole classification of wearables that are built into the clothes we wear. Some of our clothing will have sensors that attach to our mobile devices in order to provide health or safety information. In other cases our clothing will be active and able to adjust its properties based on the activities we are performing. We will have running pants that constrict in helpful places as our muscles get sore. We will have shirts that can vibrate or tap on us in some way to alert us to a condition we have pre-programmed. Some of the top clothing brands already are repackaging themselves as technology companies.

The impact of wearables will be dramatic because they will be the next step forward in allowing us to integrate with technology so that we can access information and

247

control in the world around us in real time. Mobile devices are just an early step in separating those who can use a technology well from those who do not have the ability to use the tool. As we get more devices and technology we will keep separating those who can afford higher-end gear from those who cannot, and those who invest the time to learn how to use advanced applications from those who do not. The far-reaching impacts of this separation will be seen in many areas of life. A really basic example of this already exists in sports.

Players today will invest in anything that gives them a performance advantage. When they can put on wearable clothing and devices that are intelligent and give them access to information and augment their abilities while they perform, the basic fairness of the activity comes into question. Where will we draw the line on what is legal or not in the sport? Today swimming is struggling with the properties of the fabrics that swimmers can wear because of the advantage they give. Not only does this create an unfair playing field; it also makes any record previously set an easy target to surpass. Baseball does not allow aluminum bats. Football restricts the number of players who can have a speaker in their helmet. These examples pale in comparison the advantage players will be able to gain from a wearable.

There will be a vast market for thousands of different wearables and price ranges will be wide. At the high end will be individually customized and programed devices that work much better than the lower-priced generic versions. There will be a huge aftermarket for used devices just as there is today with clothing. We will create a high amount of technology envy at this point because the performance capabilities of a person who can afford a customized set of wearables will be astronomical. This could lead to these devices being the most stolen items in the world.

Ultimately we will press against a growing addiction issue as people who are highly proficient with integrating with their wearables feel naked without them. People literally will refuse to come to work if their wearable technology is not working because they will see it as a waste of time to try and do their job or go to school if they are disconnected.

Implantable technology: We are implanting pacemakers, Cochlear implants, and eyesight enhancing devices already. We are connecting Brain Computer Interfaces (BCI) to prosthetic devices so they can be controlled as if they were natural limbs. We also have people experimenting with embedding devices under their skin to be able to use hand gesture interfaces better and to track health analytics more accurately.

For decades we have implanted equipment in order to save or prolong lives. Soon we will branch out to implanting devices in order to control technology better.

I have observed that many people alive today have an aversion to having something foreign implanted if there is not a medical reason to do so. This is not some aversion future generations will have. As the implantable technology improves on wearables' ability to control technology in real time effortlessly, early adopters will try it out. As we find out that there are no collateral health issues created we will build trust that implanting is a normal progression. Once parents have confidence they will start implanting their children at younger ages. As a child grows up knowing nothing different the stigma of integrating heavily with technology will go away.

Before most of us are ready for it soon, there will begin to be optional "upgrades" we will be able to make to our bodies. These will start as implants that help us improve a health situation we are facing like a loss of hearing, vision or memory capability. In order to "fix" these we will use implants that actually will improve a person's ability to function in these areas. The difference is that the technology we use to help people who are handicapped in some way will improve them even beyond the norm. Once we reach this point there will be people who will be motivated to augment themselves even though they do not have a physical issue. They will want an edge over the rest of humanity. There will be nothing wrong with their eyesight or hearing, yet they will choose to augment in order to gain the expanded capabilities.

One day soon we will be able to implant a BCI that allows us to control more than a prosthetic limb. These new BCIs will allow us to control a computer or mobile device just by thinking. At the speed of thought we will be able to ask and see any piece of information in an instant without anyone around us knowing. At the same time we will be able to connect with each other at the speed of thought and that will create an even more powerful Universal Mind for those who are willing to participate.

To give you a better idea of how we use the Humalogy scale in our consulting work let me extend the idea of the T1 scale even further. In the future our integration of technology will be much deeper than it is today. Technology will be an internal enhancement of our bodies as well as an external tool. Today we do not have a vocabulary to describe levels of enhancement but I am sure that will change in the future. We will need to create a vocabulary to be able to describe the various levels so we have some kind of idea what kinds of capacities people have instantly and how deeply they are augmented. To keep things consistent with the Humalogy Model, we could simply retain the T1-though-T5 paradigm:

249

The T1 level: This level would include minor modifications that include technology interfaces embedded on or in our bodies so we can control computers in some physical way. Gesture control or touching interfaces embedded in our skin to enact an action are good examples. We would include extreme use of wearable devices at this level too.

The T2 level: This level is highlighted by subdermal brain-computer interfaces so just thinking about an action can control technology. The BCI would also allow users to connect with each other in more ways than just audio and text. Instantly-shared real-time video could be delivered to a single person, a group, or the Internet in general with a single thought.

The T3 level: This level includes all the capabilities in the two before and adds light physical augmentation. This could be a light exoskeletal device to help someone walk further with less energy or a technology assist for their arms to be able to lift more weight. This will be the level where the improvements we build for handicapped people allow them to perform better than a typically-abled person. This will start a rush of people asking to have physical augmentations without there being a specific need.

The T4 level: At this level the physical augmentations are more "built-in." They extend past exoskeletal devices and now are invasive replacements of muscle and bone in order to enhance a limb or internal organ. Bionics become a more accepted modification depending on the person's need for, and use of, the augmentation.

The T5 level: At the T5 level we are bordering on a person being an android. More than half of the human body is augmented in some way and there is extensive brain and vision modification in order to access information and control devices at the speed of thought. There is a permanent connection in place with the Internet that is integrated into the person's brain. This is a level that is hard for people to accept, but one that I am very sure we will get to in four or five generations.

Now that you have a little better picture of mobile, wearable and implantable transformation let's step back and consider a few big-picture outcomes. The first is the coming disparity I mentioned between the "have" and "have-nots" in school, work, sports and life in general. It is safe to assume that people who are wealthier will have more of an ability to buy the very best wearables and implantables to augment themselves. As this trend continues we will empower the wealthy to keep widening the gap between themselves and those who cannot afford to be fully augmented. We have always had issues with class divergence at some level. There have always been the poor, the middle class, and the wealthy. Today we are worried constantly about the middle class shrinking and the fact that fewer people gather great wealth. Just imagine how much worse that may become in a world where the wealthy can augment themselves with the top wearables and implantables.

Wearables and implantables will make it harder to step away from being on the grid. The advantages we will enjoy and the frictionless nature of using tools like these to access information and other people effortlessly will be very attractive. What I am sure we will see is people who are slowly augmented increasingly throughout their lives and the further they go, the less they will want to disconnect from the grid. They will be driven to this through economic need and because they will be addicted to the results they can get when they are connected to the grid through all of their equipment. To disconnect would feel a little bit like solitary confinement and there will be many that will choose never to unplug.

Advanced mobility, with its immersive connections, will help advance the Universal mind. Early examples of Universal Mind capabilities are things like being able to exchange information between one or millions in real time. There are Websites and applications that allow people to ask questions of thousands of online strangers and get answers right away. Other sites allow people to post a request for help with a project and get it in seconds. We have built rating sites that allow us to score how others feel about services so that we can help each other make effective decisions when we spend our money. Even in our current crude state of the Universal Mind, already we are seeing significant value in being able to tap into people all over the world with a click. In each of these cases we are using mobile tools to tap into the power of our collective knowledge. As we move up to wearables and implantables we will make it easier to ask any question or share any opinion with everyone else who is connected, which may be billions of people.

The Universal Mind moves us more toward oneness as a body of people. The path from mobile to wearables to implantables enables this oneness at a technology level. It is very important to understand that we are using technology in this case to realize the God-inspired concept that we all are One. When evaluating the Godliness of technology and humankind this is a very positive outcome. As a species we are developing the ability to share our thoughts and feelings at a faster and more intimate level which, in turn is moving us closer to being One.

Inevitably we will create a real-time Universal Mind that will be a powerful force for advancing the human race. It also could have dangerous impacts on our ability to be independent, innovative and artistic. When we can reach others instantly to learn what has been done before or to solve problems for us we are not forced to be creative with solving our own problems. Too many people will rely on the Universal Mind without even engaging our own brains for a solution or idea. Even more dangerous than this will be the possibility of planet-wide group-think. The only counterbalances for group-think driving us in the wrong direction are people with the willingness to stand up as individuals and point out that the herd is headed in the wrong direction.

The progression from mobile to wearables, and then to implantables, could end with us becoming so powerful that we are not even comparable to humans who are not augmented. The word for this is "transhumanism." I will present a whole chapter on this concept later.

SUMMARY:
Mobility and the
Outboard Brain

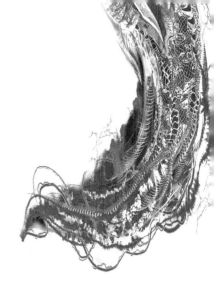

The mobile devices we carry are much more than phones; they are high-powered computing devices that are augmenting our abilities in many ways. They are acting as outboard brains because they do three things for us: store memory, solve problems, and give us access to huge amounts of information. We carry them with us all the time and we are learning quickly how to use them to make things more convenient. We are learning at faster rates because we can search for any answer to a question and get it in an instant. We are able to communicate with each other through many new channels, each of which serves a different purpose in our lives. They provide safety, help, and the ability to end any argument about a fact with a quick search. They have become so critical to many people that we would give up TV, sex, or food if forced into a decision.

We are very early in our development of these and we have a long way to go toward realizing their full potential. We will become more dependent and addicted to our devices as they provide more convenience, knowledge, and instant help from anywhere in the world. Our devices are changing us and will continue to change us even more. This is an area where we would be much better off to be very aware of the changes the devices will bring to us and to make our decisions while being more aware and mindful of the long-term impacts. Parents have been concerned over the past ten years about the impacts of such technology on younger people. They sense that mobile devices are changing how their kids operate in the world dramatically. This is just the tip of a very large iceberg that is coming our way.

The mobile device of the future will look very different than a hand-sized block of aluminum and circuitry. To get the same capabilities we have today in a smart phone we will assemble a few different wearable devices so that the functionality we seek is moved closer to our bodies and is more efficiently accessed. We will move quickly toward embedding some of these devices in our bodies so that we have even more convenient access to the Internet and all of its applications so we can operate them as if they are a part of our bodies. At that point we will need to accept mobile, wearable, and implantable technologies as a symbiotic *inboard brain.*

Humalogy Viewpoint

The balance to consider with mobility and the outboard brain is the degree to which we use our brains versus the outboard brain. This has important ramifications for many parts of our lives. There is no question that we can augment our brains by accessing Web-based services because we are seeing those results today already. What may be less intuitive, is how we will improve our lives by using mobility to tap into the Universal Mind as well. We are still well over the H side with this this potential but we will move steadily toward 0 in the coming decades. Today I put the use of our own brains versus our dependence on the outboard brain at an H3.5. because over the course of the day we still depend on our own minds to make the vast majority of our decisions and to store most of what we memorize.

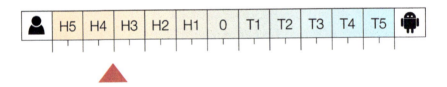

Over time I do not believe that we will move over to the T side. I would even be surprised if we move to an equal balance of using our own minds and technology when making decisions and solving problems. As I try to imagine everything that being connected in real time to the Universal mind and to the Web can do for us, I still see us only getting to an H2 balance point.

Seminal Questions

Will mobile usage lower the quality of relationships because they distract us from being present with those right in front of us?

Will mobility weaken our own coping capabilities because we become dependent on our devices for help?

How will mobility change our perceptions of privacy and what is acceptable for others to know about us?

Will the power of wearables and implantables create a dangerous schism in society between those who can afford the best equipment and those who cannot?

Will mobility drive the creation of a powerful Universal Mind that can either make us all One in a very positive way or take over our lives, leading us to dangerous levels of group-think?

Will we begin to see that our own highest nature, evolving from within, will lead us naturally to function in Oneness, connected to all by using mobile, wearable and implantable devices?

CHAPTER TEN:
The Very Real Impact of Virtuality

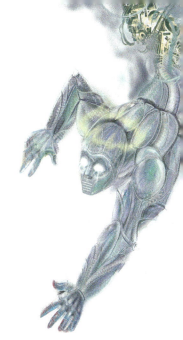

A well-balanced person of the future may be the one who can live in both the physical and virtual worlds with equal aplomb

The ability to use technology to simulate things that are not real in a physical sense has large ramifications and potential. Virtuality (my created word for this chapter) is being applied in order to entertain, teach, help us communicate, and to visualize worlds that are not real in a tactile sense. Virtuality is the ability to use software and hardware to simulate something or someplace in order to make it feel as real as possible to a user. This capability cannot only replace our physical world with a complete virtual world. It also can augment and extend what we see and experience in the physical world. We have long had our imaginations as our tool to "see" a kind of alternative world; virtuality stimulates our imaginations by giving us realistic visual stimuli that we accept as real while we interact with them.

Virtuality may be a method for us to learn how to use our imagination at much higher levels because we are practicing the suspension of the reality we see right in front of us continually in order to accept a different reality. The reality each of us sees is not the actual Reality with a capital "R;" it is the reality we create and are willing to accept for ourselves. For hundreds of years people would not have believed that humans flying would have been reality. A hundred years ago people would not have believed that it would be reality that we would have tools as powerful as the Internet and mobile devices.

God may be helping us accept that there is a spiritual reality (realm) that many cannot accept today for the same reasons we have struggled in the past – spiritual energy cannot be seen. For many, they cannot accept what they cannot see, touch and feel. Virtual reality may be conditioning us to accept things that are not "real" to our current senses. In that way we will better be able to accept that there is a spiritual level that we cannot see physically. We can suspend reality to play a virtual game because it uses the senses of sight and hearing so, even though the content might be otherworldly, we can at least accept it as real for a time. Accepting a spiritual reality where our souls communicate at an energy level is more difficult to accept because we cannot use our main five senses to verify its reality. We hear stories about people who can sense something happening miles away but we do not know how to explain the connection outside of our known realm.

For some people the concept of believing in anything that is not physical and right in front of our faces is foreign, strange, and somehow wrong. For others, there is acceptance that what we see is only one realm of many that include things we cannot see, much like we know that infrared light exists even though we cannot see it. Virtuality is definitely a technological tool that is easy to accept if people use it. For those who have never experienced a virtual world or virtual reality, the concept normally is just too far out. Those unfamiliar with the experience are mystified that anyone would spend real money on virtual goods that can only be used in an artificial world. They see no value in participating online as an avatar when the tangible world has plenty of things to keep our attention. Possibly the reason for this is because technology-driven virtual worlds are scary and unknown to them. For a person who has never been conditioned to use technology to operate in a virtual world, the first excursion might seem massively disorienting. We can probably generalize and say that younger people today grow up playing digital games that eventually take them into fully immersive virtual worlds. Older people did not grow up with this and, in many cases, have never even experienced a game of this type. For Traditionalists, virtual goods and worlds are one step too far in using technology to replace the tangible world for which they have no pent-up desire.

Regardless of our level of comfort with things virtual it is, and will have, a huge impact on how we live in the tangible world. The ability to create simulations is powerful because it allows us to experience things either that do not exist in the tangible world, are too dangerous to experience, or are not possible for us to experience physically. Simulated worlds also give us a chance to exist in that world in a different way than we do in the tangible world. We can have an avatar (a character or player) that looks different than we do and has alternative capabilities that we do not have. Being able to be a different person (or even a different species) can have interesting psychological impacts on the user. Both of these possibilities can be very seductive as a alternative or supplement to our regular lives.

It is important to note that physically creating virtual capabilities has been difficult because we have been limited by the hardware and software itself. In order to simulate something as lifelike to us it takes a ton of computing power to render the graphics fast enough to seem like they are lifelike. It also takes immense computing power to calculate what the user is doing and then adjust the virtual feedback so that the user's actions are in context. If you are in a virtual environment and move your head or change your perspective, the hardware and software have to adjust everything you are seeing fast enough that there is not a perceptual lag in your brain. If there is, that lag will make you dizzy and eventually create motion sickness at some level. Add multiple users to the virtual world all doing unique activities and you can see how a heavy amount of computing power might be needed to keep the world running smoothly.

The larger the virtual world is, the more we can move around within it and the more fascinating it becomes. Also the larger it is, the more the hardware and software must struggle to generate the virtual world because the potential variations multiply geometrically. The more user-driven capability we have in the virtual world to interact, the more variables must be accounted for in the programming. This is the reason that the highly realistic games we have today need powerful hardware and software systems to drive them. Back in the early 80s we had games like Pong: a square that moved around the screen with two lines as paddles and that was about all our computing power could handle. Then we progressed to Pac-Man and we had chunky pixels that simulated a maze and the bad guys tried to eat our character. Again at that time, this was a difficult capability to provide because of all the options and real-time graphics that had to be generated. These examples are crude compared to what we have today in the gaming world and yet we are still limited by what the hardware and software will allow. We have to cut down our first-person perspective to something much less than a 360-degree view for example and we have to limit capabilities just to what we can render with our current graphics engines.

We are not limited by our imaginations or the valuable uses for virtuality. We are limited by the constraints of hardware devices and software platforms. This is important to understand because computers are getting smarter and more powerful all the time which means that our ability to create virtual tools will explode right along with the power curve. We have improved our ability to virtualize environments tremendously over the past twenty years and there will be no slowing down for the next twenty. Now we have invented 3D, 360-degree cameras that can record a "movie" or environment to be used with a virtual reality headset so a user can further suspend reality by choosing to watch a scene from any angle. If you think scary movies are frightening today, imagine when you participate inside them with a 360-degree viewpoint where the monster can approach you from any side.

We are on our way to full realism where what we are able to see virtually will be almost as real as what we see naturally. The closer we get to this ideal state the deeper we will be able to lose ourselves into virtual reality. This is both exciting and a bit scary, considering the consequences. Think about the Holodeck from Star Trek if you are a fan. This is a room where, once a person enters, they participate in a scenario that their senses will tell them is completely real. Think of the possibilities; we could interact with a long-dead relative. We would be able to simulate running in a race and feel as if we were really at the location among hundreds of other racers. We could be trained to do any physical activity by practicing with virtual players or trainers. We could practice going on dates and making sure our behavior is suave and debonair. We could even relive scenes from our past if we were willing to spend enough time loading the information about what happened into the system.

We are in the very early days of high-quality virtuality, meaning that we can produce it fairly well, but only with huge computing and programming resources. When there is enough motivation, like in simulating control of an airplane in order to train pilots, we can mimic reality very well. Note that we have not been quite as motivated to simulate what it is like to drive a car which help new drivers practice without the high consequences of accidents. As a society we want pilots to be perfect with their capabilities, yet we do not require the same type of training for our teenage drivers.

As the ratio of price-topower for computing capability keeps improving and we invent better visual/auditory devices to deliver the simulations, we will see many new uses for virtual reality emerge as part of a natural progression. Today there are four major areas of virtuality that are growing quickly:

1. Virtual worlds that entertain us

2. Virtual reality that can help us train to do a task

3. Augmented reality that helps us to be more productive by augmenting tasks we do in the real world

4. Augmented reality that helps us digest more information about the world around us

Virtual Worlds

This is the creation of an entire simulated world using software. Typically the user is able to move around (walk or fly) through this world and it is large enough that there are many different places to navigate. This world could be a simulation of our actual world so that someone can visit a place without physically being there or it could be an alternative world with very different conditions and properties than our tangible world.

We have developed programming platforms that now allow developers to build fairly realistic environments and to allow a person (or player) to navigate through the world to experience whatever the goal of the virtual world might be. When the software environment is matched with hardware devices that facilitate the navigation, the user will have multiple options for controlling their avatar on screen. This means the avatar can run, jump, roll, and climb through the pressing of buttons or maneuvering a joystick of some kind. Add virtual reality goggles and the world can move from being a 2D environment on a flat screen to 360-degree view that increases the realism.

Examples of virtual worlds range from applications such as Second Life to the many game worlds that have been developed over the years. Second Life was built to be a full alternate world with its own sets of rules for behavior. The gaming environments vary widely and are geared for different ages and tastes. Over the past fifteen years the game worlds have continued to become more photorealistic. They can be filled with other players (avatars for real people) and non-sentient characters who are little more than props in the game.

Virtual game worlds have grown in sophistication and realism over the past fifteen years and have benefitted from the hardware and operating systems that have provided additional power. For the players of these games, part of the attraction is the ability to lose themselves in these worlds as if they are real. This drives game developers constantly to find new ways to provide realism for the players.

Most game worlds are mission-based in that they are designed to guide a player (user) through a preprogrammed set of tasks and stages in order to accomplish a goal, win the game or defeat on enemy. There is normally something that the player (or citizen) is guided to achieve within the game world depending on what kind of environment is created, however there are some virtual worlds that are created simply to allow participants to interact in some way. At one end of the spectrum there are educational games that are built specifically to simulate something in the tangible world for the purposes of providing lessons that a student can learn by playing the game. The activities are designed to imprint information or train the player/student with some physical skill that can then be replicated in the tangible world. This class of games is sometimes referred to as "edutainment" and, in some cases, is based on virtualizing a situation or real life to some degree. These games have proven to be highly effective and some studies have shown that they are 20% better at teaching curriculum (according to dora.dmu.ac.uk). The reasons for this improvement range from the the ability of a student to repeat the learning over and over which cannot be done in the classroom, to the simple fact that young people can sometimes focus more intently on the activities on a screen when compared to a classroom. Rarely do students get addicted to this type of virtual environment because there is nothing in them that cause ongoing psychological attachment.

There is a new class of game worlds that is pretty exciting because the players are doing activities that aggregate their brainpower in order to solve problems or improve something in the tangible world. Think of these games as a step toward the Universal Mind in that they are built specifically to achieve a goal, from getting work done like indexing millions of pictures to solving real-world crimes. The players may or may not know that they are being sourced for their collective brainpower. We have already had a number of groups experimenting with this type of virtual reality and I believe strongly that we will see more online games built for this purpose. Some day these game environments will harness crowd dynamics to do many positive things in the world.

At the other end of the spectrum there are virtual game worlds that have no redeeming value other than to entertain people through simulated violence. There is no small market for these, of course, given that billions of dollars are spent on first-person shooter games that people want to believe are harmless fun when they are not harmless for all players. Some of the most popular games are based on war themes or where the specific mission is to kill the enemy presented onscreen. While there are players who enjoy these games and are not desensitized to the blood and killing, there are others who are impacted psychologically and carry that damage into the physical world.

I am amazed that parents are so lax in keeping their kids away from war-based and first-person shooter games or virtual worlds like Grand Theft Auto, which combines the worst aspects of humanity into a virtual world. The scary thing is that these game worlds will keep getting more realistic and graphic and, as they do, people will be able to lose themselves inside these games and suspend reality for hours at a time. When they do emerge and step back into the physical world, they will not be able to avoid the darkness impacting their minds and emotional states. This combining of realism, entertainment, and violence will be a serious problem in society and we will be forced to control it better than we are today. Ratings on a box are not going to get us where we need to go.

There also are virtual world games that do not have a goal of violence. The Sim City series of games, for instance, provides a very different goal for their worlds because the player is responsible for building something complicated and seeing how it functions over time. There are many simulated sports worlds that allow the player to play football, soccer, tennis, and other sports. There are simulated music games that allow a player to be part of a band and play an instrument. As these games improve their realism they will begin to be training tools that can help young players advance their real-world capabilities and will be used as training aids. I see virtualized games like these being a very healthy thing as long as people do not substitute them for getting exercise in the real world too much.

Second Life provided a completely simulated and non-guided world. This was one of the most highly-used virtual worlds that was not a game world. There was no mission or competition built into it. Second Life simply is an alternative version of our real world, provided virtually. In other words it simulates the tangible world pretty closely and does not provide any structured goal to achieve. People have avatars that represent themselves on the screen. They dress their avatars in whatever outfits they want build them to be any human (or non-human in some cases) shape, color or nationality. A member of Second Life can buy property, build whatever home or building they can afford, can build it themselves if they learn how to construct virtually. They could build virtual products and sell them in stores. They could talk to each other and interact physically with each other and do physical activities, including virtual sex, fighting, walking, running, or flying.

I once owned property in Second Life and had built an amphitheater and a few other buildings, including a multi-floor mall of sorts for virtual goods. We experimented with playing videos of me speaking so that an avatar could walk in and watch me present some technology idea while they sat in our location. We paid another person to buy the property from the original owner and then paid Second Life monthly

to "own" the property. We developed the property over time and it had lots of interesting architecture and capabilities. One of the cool things about a virtual world is that we can build capabilities into our buildings or properties that are not easy to build in the tangible world. Also we can build out locations much faster and less expensively than a building in the tangible world. One day I put a space suit on so that I could fly (my avatar) up into the sky. I heard that some people were building floating buildings up in the airspace above the terrestrial buildings which did not seem right to me. I assumed I owned all airspace above my building as far as Second Life could possibly allow airspace to be. After a 3000-foot virtual climb I found a very cool floating nightclub above my building. Obviously this was not visible from the ground. There was not a door into this club so I guessed it was some kind of private affair that was invite-only. I used a special technique to slide my avatar into the building and, lo and behold, it was a strip joint, or at least the avatars at the club were sans clothing. I had to ask them to move somewhere else. This is a good example of how a virtual world can have different properties than the real world and, at the same time, present some of the same moral issues.

People who have invested time and/or money in Second Life believe in the possibilities of a virtual world where many of our physical world activities can be simulated in order to become free of physical barriers. The great benefit of a simulated world like Second Life is the ability to do a number of activities from real life without leaving home. The bottom line is that, although Second Life has not continued to enjoy the popularity it once had, it definitely pioneered the way for alternative worlds that surely will come in the future.

During the hey day of Second Life I was fascinated by the church that Life Church in Oklahoma City built in the virtual environment. The experience they created was more realistic and engaging than what could be delivered on a Web page or through a television. An avatar would walk into the church and was greeted by another avatar. The message was delivered by video on a big screen in a room that looked a lot like the actual building Life Church had in the physical world. There was a lobby area that simulated a lot of the services they provide in the physical world. When I sat in the room to watch the message I was surrounded by other avatars that all had real people behind them. The only aspect that was odd, other than attending church virtually, were the bizarre costumes people sometime adopt in Second Life. Sitting next to a person dressed as a skunk, a punk rocker, and super hero was a bit distracting. With that said, it was a new way to hear the message of God and I felt like I was seeing the future when I participated in services.

Virtual worlds transport us into a different realm from whatever physical situation we are in at that moment. They also provide ways for us to modify those worlds to improve capabilities over our physical world. For instance we can move anywhere in the virtual world in an instant. We can appear any way we choose to look (by changing our avatar). This ability does cause a bit of dissonance between the person we are meeting online and how they look versus what the real person behind the avatar might look like. From a spiritual viewpoint I believe that we should look past the physical appearance of people anyway so I never got too strung out about the costumes people adopt online. I have always believed that what people chose to where or how they chose to mutate their avatar was a message they were trying to send and it needed to be noted, but not overly evaluated. We can give the world properties we don't have in the physical world (the way the geography looks). And we can build fantastic buildings and float them in the sky in ways we cannot today in our physical world. In a virtual world we can float and deliver information anywhere in our vision, triggered by any behavior or location. In short, we can build much prettier, fantastic, and productive worlds in the virtual space. Minecraft will likely go down in history as the largest and earliest example of the ability to have millions of people assemble an online world. It has already stood the test of time, having been active for many years now.

Virtual Reality

Virtual reality (VR) often refers to a more particular application of virtualizing in that often it is built for a specific task. Instead of creating a full world, VR can be used to simulate a very specific task. For example, how to repair a virtual engine without ever touching a real engine. VR also can be used to simulate a real-world location by using actual video taken from the locale. The VR equipment allows the user to be in the location and have a 360-degree view of what it is like in person. We also tend to use the VR term more when talking about the hardware devices that are used to display a 360-degree ability to exist in a virtual space (e.g. VR goggles, glasses or headsets). Examples today of VR devices are the Oculus Rift or Samsung's Gear VR.

The military has been a pioneer with using virtual reality to train troops in the use of various weapon systems and dealing with different scenarios they might face in battle. This is a perfect example of the need for virtual reality because there are many aspects of warfare that simply cannot be taught in the real world without creating dangerous and life-threatening training scenarios. There is a growing list of other applications that are being built for training in areas like health care, manufacturing, and many types of repair work. It is much less dangerous and expensive to teach someone to do complicated skills in a VR environment than in real life with real consequences.

VR as a way to train ourselves to do a specific task already is helping us in meaningful ways. I already mentioned the example of the airline pilot that can get many hours of "simulations time" under his/her belt in order to feel more confident in taking on the responsibility of hundreds of lives in the air. In the health care environment VR is used to help a surgeon practice and experience what a surgery will be like without having to learn on a live patient. These are both early examples of what will be a huge part of our lives moving forward. As costs come down and the technology to program and deliver VR gets simpler to develop, we will start seeing VR training for hundreds of applications. Anything that requires us to use a complicated series of motor skills and make multiple decisions at the same time will be target-rich for VR training. Tasks that we used to learn by rote practice (playing the piano, learning to swing a golf club, or knitting) soon be learned or improved by putting on VR glasses and loading a training program.

VR training/learning will speed up our ability to gain skills and in some cases will provide a level of safety that we have never had before. No more dangerous practices with a learning process where physical harm to the user or equipment is the price to pay in training accidents. A month with the right kind of VR training system will be like a year of practicing the real process. Kids who never would have had the opportunity to train for something that requires expensive equipment, or who did not have access to a teacher, will be able to use VR training whenever they want. After completing the VR training they will have the measurable proof that they can apply what they learned in the tangible world if they choose.

VR not only will help us with careers built around physical tasks; they also will have benefits for knowledge-based careers. If we work in a career that is more white collar in nature, VR will allow us to extend ourselves geographically by providing collaborative environments and communication that makes it feel like we are in one room with people who are very far away. Often this is called "telepresence" today, which is an example of a capability that makes us more productive by saving travel and expense. The ability to meet in a simulated room with people from all over the world allows us to exchange ideas and information that would have been done through phone- and text-based conversations in the past. Although an email trail can exchange information, it is difficult to get the full context of how someone feels or build any kind of relationship without seeing their face and body language.

Virtual environments can be very entertaining and practical and, at the same time, can be a place where a person can feel more comfortable than in the physical world. That is where the danger starts. If we can psychologically "visit" a virtual world and use it for the reason it was created, thereby keeping a separation in our minds between what is real and what is simulated, we can step back into the real world un-phased. To the extent that someone completely dispenses with reality while in a virtual space, they have taken the first step toward acting out simulated learned behaviors in the physical world. And if those behaviors are negative the virtual reality has just trained a person to step into the dark side in the tangible world. Since our past history shows that a certain percentage of humanity slips into the darkness, virtual worlds may become a powerful breeding ground for warping people psychologically.

Let me explain this in more detail by looking at two different activities that can be simulated in a virtual world with dangerous consequences. The first are "Synthetic Relationships" and the second are "Synthetic Experiences." I am going to use the word "synthetic" to describe these behavioral aspects of a virtual world because it describes well how it feels to us. Let me illustrate by using a simple hug. There is a difference between hugging a person, hugging a doll or your avatar hugging another avatar. When we hug another person there is a sense of touch, smell and an energy exchanged. We feel the comfort of their body next to ours and there is a trust and vulnerability that come with physical contact that is that close. There is an energy that is exchanged that we might subconsciously feel, which is a very real exchange of love or friendship. There are side hugs, full-frontal hugs and uncomfortably long and hard hugs, and each one sends a very different message at many levels: emotional, physical, and psychological. A hug can make us cry or make us happy. It can bond two people in a deeper way or be incredibly uncomfortable. It is a "real" experience no matter what.

What about hugging a big life-sized doll? There is the physical sensation of putting your arms around something but, because it is inanimate, it cannot really hug you back and even though you have the physical sense of embracing something human- like, you do not get any energy exchange. It might feel a little bit good to you but only as a poor substitute for the real thing. Now let's go to the next level and talk about a hug between avatars. Now you have the intellectual stimulation that your avatar is merging with another on the screen but there is no physical sensation. Actually, you cannot really be sure who is controlling the avatar you are hugging. This synthetic hug might be better than nothing but it is not a real hug. It lacks the physical experience and the energy exchange. There is much less nonverbal information exchanged as well. You do not gain the most valuable benefits of the physical comfort and meaning in a real hug. The synthetic hug is a symbol and it might help us feel more comfortable hugging people in the physical world because we have practiced it over and over in the virtual world, but it is not the same.

With that basic example let's look at synthetic relationships with the same lens. We have already discussed online relationships in detail so, in this case, I want to limit our definition of synthetic relationships to those led in a virtual environment. Through a business lens this relationship involves working with people within a virtual office space. I am not talking about video conferencing or Skyping someone because, although this is now a real communication capability, there is a real person on the other end and not an avatar. During its heyday Second Life had a number of companies that built virtual offices and attempted to hold meetings in this way as a method of cutting costs and raising collaboration levels. For the most part it was business etiquette to build your avatar to resemble, at least within reason, what you looked like in the physical world. Meetings would be held in what looked generally like a conference room. Many other companies created their own virtual office environments with various types of unique functionality like being able to open up windows for documents or videos, or the ability to put a picture of your real face on your avatar. Some of these companies still exist and are used in a limited way in the business world.

Having a personal synthetic relationship is a step up from an email or text message relationship because you have a visually simulated "person" communicating with you and there is a physical dimension in that your avatars can perform actions like human bodies might. With an avatar you can wave at someone, hug them, talk to them, smile at them or even take off your clothes. This allows a person to "feel" like the relationship is more real than just a more abstracted online communication relationship. This allows a person to fantasize about what the other person looks like and to humanize the relationship to some level. As with anything else human, there can be positive or negative dynamics in a synthetic relationship, depending on the circumstances. It could be a fulfilling, friendly relationship with an exchange of ideas and information that fulfills both parties and keeps them from being lonely in the tangible world. This is especially true for a person who, in the tangible world, is confined in some way. Or the synthetic relationship could be based on the parties trying to replace the affections of a spouse or play-acting as a different gender in order to deceive or fulfill a latent desire.

We are in the very early days of this capability. As VR goggles bring higher quality and as more detailed and functional virtual worlds are built, the ability to build synthetic relationships that have very real meaning will grow. No doubt that one day the line between real and synthetic relationships will blur. This may seem crazy to people who have never experienced this kind of relationship but the generations who grow up with friends who they only know through online connections will feel increasingly comfortable with having both offline and synthetic relationships and not distinguishing too heavily between the two.

I worry about emotionally dysfunctional people who have all their close relationships in synthetic ways because that is less scary, or because they will not get arrested for fulfilling whatever relationship need they are looking to fulfill. The porn industry has been on a steady march toward doing anything it can to deliver a real sexual experience over the wire, which has led to every kind of strange device that can be controlled by a remote party. People who just want to make money any way they can already prey on lonely people with spam emails from "lonely girls" looking for a man. Imagine when that girl/avatar can walk up to a person in a virtual space and say hello. The experience of touching someone will be reproduced synthetically and there will be a lot more damage to us than there will be benefit.

On the more positive side, there are amazing experiences we will be able to give people such as allowing a handicapped person to walk normally in a virtual space with a fully-functional body. We will be able to provide people with a synthetic experience of doing something they could never afford to do in real life like see another part of the world as if they are there walking around. We will be able to let people experience doing a repair step-by-step at whatever pace they want as if they are there in person and are using the right tools. From anywhere in the world where a person has access to virtuality, they will be able to see what it is like to rebuild a car engine as if they are doing it with their own hands, and that will be extremely fun and educational.

There will be amazingly positive attributes of synthetic relationships, experiences, and virtuality in general. At the point when our capabilities mature to be even more realistic and virtual worlds grow to be expansive and inexpensive to use, our young generations who have no fear of technology naturally will adapt. They will get better and better at using these new tools to accomplish thousands of new tasks and, in doing so, will become more advanced, knowledgeable, and skilled than people who do not participate in virtual worlds. We would be naïve to believe this will not happen over the next few decades.

Augmented Reality (AR)

This variation on virtuality refers to using a visual device to enhance our physical world by providing a virtual overlay of something we see. AR software recognizes the object we see and injects pictures or data to augment what is in our vision. If we are looking at a city block our AR device could be adding a building to be built next year where there is an empty lot today. If we were walking through a graveyard our AR glasses could be adding information that floats above the headstones about the person buried there. AR simply is adding information or visuals to anything in our current field of vision.

A simple example is a heads-up display (HUD) on a vehicle. A more sophisticated example is an AR application that walks a mechanic through doing a repair by adding step-by-step instructions into their vision field while they look at an engine. Today there are many AR applications that use our mobile phones. These can help us visualize where train stations are in relation to where we are standing, what is in the building we are looking at, the names of the mountains in front of us or physically where all of the restaurants are around us.

In order for AR to work there must be some kind of device that allows us to see what is in front of us while projecting additional data or graphics. Keep an eye out for AR applications blossoming around you because there are many valuable uses for augmenting what is in our view at any moment.

Augmented reality has huge potential in many areas of our lives and already is available for use in a growing number of applications. There are examples of AR capability being delivered in mobile apps in ways we might not recognize as AR. Anytime we can look at our mobile device and get additional geographic information overlaid on our screen we are experiencing AR. The concept of AR was moved another step forward recently with the work Google was doing with Glass and that Microsoft is doing with their HoloLens. Although Google is done with their first generation of Glass at this point, they have helped paint a picture of what a heads-up display could do in order to enhance the information about whatever we see. Hopefully Microsoft and others will pick up the gauntlet and develop more valuable capabilities.

For years science fiction movies and books have provided characters with heads-up displays (HUDs) that inject additional information into the users line of vision. The U.S. Air Force provides this for pilots of planes and helicopters today. Most first-person games provide a simulation of a HUD for the player. This concept makes all kinds of sense because we really get very little information just from looking at someone or something with our eyes. We look at a person and we can tell if we know them or not. We can make some assumptions by how they are dressed or act but other than that, we glean little from experiencing a person only visually. With full AR we could look at a person and see a text-based floating data stream that provides much more information on who they are and what we might want to know about them. We can look at a car and see what year it was made and possibly any other fields of data that exist publicly about the vehicle. We can look at a restaurant and see immediately how full it is, the ratings from other diners and what the menu has on it tonight. You get the idea. We will be able to look at just about anything and have our system augment what we are looking at with loads of ancillary data that might be helpful.

We can also use AR to envision something that does not exist in the physical world. We could "see" what a room would look like painted another color or with different furniture, what a block of downtown would look like with a new park in it or what a person would look like with a different outfit on. If you have been paying attention to the news, all of these capabilities exist today.

Combine some of the areas I have already shared and imagine a world in which we have implantable devices to control technology through a brain-computer interface and an optical projection capability that pushes visual information into our eyes without glasses. We will be able just to think a query and the information will come to our vision in context of what see. As you probably can imagine there are many industries that will be impacted by this capability. AR at that level will help us "see" more about the world around us by adding loads of data into our field of vision and we will become very adept at integrating both what is real and virtual. This is not just science fiction; these are exactly the kinds of capabilities our grandkids will see as normal. And this is the reason we need to think about what the impact of virtuality will be on humanity because our capacity to slide toward being like cyborgs is a very real possibility.

As we develop more advanced skills with using VR or AR in our personal lives and careers they will progress rapidly forward with what we can achieve. We will use simulated environments to learn faster, to experience situations earlier, and to replay them many times over so that we are comfortable with the solutions. As we do this there is more potential for a separation between the people who are augmented and those who are not. There may also be a corresponding diminishment of our ability to build deep and meaningful relationships if we are too distracted by our immersive connection to technology. Earlier in history an architect who used computer-aided design was clearly separated from one that still created blueprints by hand on a table. A furniture builder who used power tools and nail guns could build furniture faster than one using hand tools. In the same way a well-used mobile device can separate one executive from another in their ability to be productive. The next big horizon for separation will be the ability to use augmented reality and virtual tools in order to be more efficient.

The Negative Impacts of Virtuality

Along with the positives of virtuality will come the negatives; I have mentioned a few of these already. So that we do not go blindly into the future of virtuality here are more dangerous consequences I would like for us all to avoid:

- One never knows who is really controlling an avatar in a virtual space and that is not the only deception that happens. If we are working in a virtual world where there is a simulation of a location or experience, we really will have no way of knowing if it is accurately duplicated digitally. A surgeon trainee could be doing a simulated surgery and the virtual training program could have a flaw in training on how the virtual body reacts to a procedure. Until this is discovered we will be training surgeons to do something that will not work in real life.

Or we could be in a simulation of a location we are interested in visiting and the visual experience is quite overinflated. I booked a hotel in Switzerland once that had a picture of a beautiful lake behind it. When we got there it was in the middle of a city and there was no lake to be found. As "virtuality" becomes more mainstream we will suffer many versions of this problem because companies will be motivated to alter reality in order to make their location seem better than it is. We may eventually have to pass laws to protect people from unscrupulous businesses that fail to deliver truth in advertising through virtual marketing.

• We are just coming to understand the invisible flows of energy that happen between people or things in this world. This dimension of energy flow is critical to us as human beings because it fulfills, warns, and informs us at a sensitive intuitive level. This is why we say things like, "the vibe in the room" or, "I got the impression." It is impossible to get this level of energy exchange in a virtual experience fully.

Although virtuality will provide the ability for us to experience different kinds of simulated situations from the peace and quiet of our home or office it will not replace the energy that is exchanged or provided in the physical world. The key word here is "simulation" because what we experience in a virtual world may feel real if we can expend reality for a while, yet it will not fully feed our soul or heart with the energy that gets exchanged between real people and real in-person experiences.

• As with any dependence on technology, sometimes there is an accompanying weakening of real-world skills. In the virtual space this could happen through an over-dependence on synthetic relationships for fulfillment. There can also be an out-of-balance desire to live in a virtual world where one can have ultimate control and always be the hero. Psychologically a person will find themselves on a slippery slope if they gain the majority of their positive feelings from simulations. As with any other addiction, the unhealthy connection to virtuality for fulfillment will leave people empty inside.

I believe we are seeing a bit of this already with people who spend too much time in game environments. Normally they will not admit that eight to ten hours a day of playing computer games is unhealthy. As virtual worlds become more realistic and not game-based there will be a segment of society who falls in love with their virtual capabilities. And just like a gamer, a person who yearns to use virtual technology to escape who they really are might also find themselves deeply connected to a virtual world that ultimately will not fulfill them at all.

• It's frightening to consider what hackers will be able to do in virtual spaces. They will be able to steal avatar identities and be imposters to unknowing people who will give up any kind of secret because they believe they are dealing with a trusted avatar. They will be able to alter virtual worlds to suit their schemes and will be able to steal virtual goods and resell them to others just like criminals in the tangible world. And hackers in the virtual world can do any of their dirty work in broad daylight and from anywhere in the world. I expect that virtual hacking will be a new frontier very soon.

• As virtual uses expand there will be a cognitive dissonance between their two worlds for some people which will lead to all sorts of psychological and emotional issues. They will, quite literally, be two different people in their two different worlds and, as the virtual becomes more lifelike, this dissonance will become painful. Already this is happening to a small percentage of heavy gamers who would rather live in their game worlds than in the physical world. They feel comfortable gaming and are very highly thought of by others in their chosen virtual world, but when they go to school they are shunned by fellow students for their behavior, clothes, or appearance. Look for this to happen to businesspeople, housewives, teachers and many other mainstream people who have a reason to spend a lot of time in a virtual space.

The Future of Virtuality

When I turn on my high beams to see what all things virtual might look like in the future I gaze at an intriguing picture. There will be many everyday uses of virtuality that would seem very foreign to people today. That is unless they have been reading lots of science fiction, however. These writers often do a wonderful job of projecting not only what new inventions we might have; they also tell stories about how they are woven into our lives. Virtualization is a common component of many books I have read and, in most cases, the practical uses resonate as being very probable to me. I will start with future trends that are easy to predict and then move to the more thought-provoking speculation.

An easy prediction is that we will see constant improvements to the equipment and software that help us simulate tasks or environments. When we see anything visual on the screen we want it to portray what we would see in real life as accurately as possible. That means we want our virtual worlds to be a very close approximation of the physical world so it is easy to translate how we would act in the real world onto the virtual world. We know how to use our hands already so being able to use them to gesture control anything on the screen would make sense. We are used to interacting with physical things in the real world like a paint brush, so that is why graphics programs use them on screen tools for us to change the color of something. The more lifelike the tools are, the more comfortable we are using them in cyberspace.

We also want our augmented reality to inject additional data and graphics into our physical world as seamlessly as possible. What this means is that when we walk down the road, drive a vehicle, or are busy working, we want a constant stream of meaningful data subtly visible in our peripheral vision. This increases the amount of real time information we have access to dramatically. What is obvious to me is that we will perfect augmenting the scope of the vision we have today one day by adding lots of information that is pertinent to the activity we are doing. When we watch sports there will be a constant waterfall of information about the game in our vision if we choose this. While we work there will be a constant dashboard of information about the field we work in streaming into the edges of our vision. When we walk up to people there will be a column if profile information in our vision that comes along with their proximity. Augmented reality will seem completely normal to the child that grows up knowing nothing else. They will feel blinded without it, quite literally.

Japan is making more progress than other parts of the world toward creating simulated humans either with robots or online avatars. They seem to be intrigued with mimicking rock stars or actors; the more lifelike, the better. Because they are forerunners we can look at how they are relating to virtual people and see interesting trends. Although they are striving to create accurate looking avatars they also have started to experiment with stylizing their avatars with unrealistic physical attributes. What has been reported is that they like their avatars to be partly lifelike. They want to have clues that the virtual person is not real so that somewhere in the back of their mind they can be aware that what they are looking at actually is a creation. If we make a virtual person so real that people cannot tell the difference, some people start to get uncomfortable. The solution is to add some kind of unrealistic physical characteristic – longer legs, and smaller waist, bigger eyes – anything that is not humanly possible in order to give the clue that the avatar is manufactured.

As we sit here today most people are a little disconcerted by the thought of interacting with virtual beings or robots. Although understandable because they have not done this throughout their lives, there also was a time when they would not have imagined they would spend hours a day interfacing with a small handheld screen. We thought it was science fiction that a small magic box could entertain us, inform us and solve our problems. Mobility has become a very normal part of our lives relatively quickly. Virtuality will creep up on us in the same way. There is no question that interacting in virtual worlds, learning to do things through virtual reality, or wearing glasses that enhance the quantity of information being provided to our eyes will feel very strange to most people who have trouble adjusting their perspectives after decades of doing things in a physical way. They trust what they can see that they perceive as real, they trust what they can touch and they trust a person they can interact with face-to-face. It will be hard for them to trust what they do not perceive as "real."

With that said, technology and the benefits it delivers win most of us over at some point. There is a psychological switch that will get flipped in people when we let go of believing we can only trust what is "real" or what is human. With this change of mindset, we will accept that participating in a virtual world has benefits and that owning virtual goods makes sense because they also have value even if they are not tangible in a traditional sense. We will come to love augmented reality and its capability to provide much more data about the world around us. We will even begin to trust the benefits of synthetic relationships and experiences. There will always be the purists who reject the concept of integrating the virtual into our lives, and that is okay. The majority of people in the world, however, will come to accept virtuality as a normal part of our days.

When we get to the point that a majority of the world uses virtual tools, a very negative dynamic will exhibit itself: the growing discomfort for some people with the physical world. The big difference between the virtual world and our physical world is that we have more control over the virtual world, which means it is more idealized. Our avatar will never grow old, gain weight, get sick or feel pain. In the virtual world plumbing never breaks, the weather does not destroy our homes and just about anything we want to do is accomplished instantly. The tangible world is much slower and fraught with problems. As time goes on we will come to love the virtual and augmented worlds because they are cleaner, faster, and the problems are so much easier to solve, not to mention there will be very few problems in the virtual world. Our real worlds will always be filled with frustrations, pain, and friction. This might sound like a negative view of the tangible world, but I do not want to give that impression. We do need to understand that, by comparison, the tangible world will be a very frustrating place to get things done compared to the virtual world where we work or play.

Let's take a step even further into the future and look at how virtual people (not avatars, but fully-functioning virtual bots) might interact with us in virtual spaces.

We will develop intelligent avatars that can act as humans in a virtual world. They will be easy to spot in the early days because they will be a bit stilted or robotic acting and sounding in the early days. Over time we will improve our artificial intelligence capabilities and will be able to use these to simulate human behavior in an avatar. At some point it will get harder to know whether we are interacting with a real human or an intelligent avatar in a virtual space. When this happens we will cross another line that will cause dissonance problems.

We will be able to design intelligent avatars that have specific archetypical properties. We will be able to program motherly avatars, helpful service avatars, sexy avatars or wise mentoring avatars. In each case they will fulfill some role needed in our personal or work lives. As they get more realistic we will develop more real attachments to them because of the value they offer in our lives. Being attached to a virtual world because it is efficient is one thing; being deeply connected emotionally to an artificial being is very different. The attachment to idealized avatars will be exceedingly dangerous because real people never will be able to compare to a creatively built persona. With the ability to check off a series of choices for how I wanted a female avatar to behave, I configure a virtual persona who would treat me in just the way I want to be treated. If she did, not all I need to do is adjust her programming and she would improve instantly. Alas, this task in real life is not so easy.

It is reasonable to project that an idealized female avatar developed by a man will be able to interact exactly as that man would choose. It will look and act exactly as the creator chooses. If the avatar looks and sounds very realistic it becomes easier to have a relationship with the avatar that substitutes for a real-world relationship. Obviously we have many fictional stories being written already about people becoming attached to robots because of the potential to have them be exactly what we want them to be. Long before we are able to build a fully functional robot that allows us to suspend reality enough to be in love with it we will be able to program an online avatar that simulates a true-to-life emotional connection with us. This is because the avatar will be able to communicate with us in a very "human" way. Within a virtual world the avatar will be able to act and move just as we choose. This will be harder to create for a physical robot and will be much simpler when it is just software we are customizing.

Now imagine stepping into a virtual world where we interact with avatars and we are unsure if there is a real person controlling the avatar. Another possibility will be that a real person has put their avatar on automatic so it acts like the person you think you are interacting with even though they are sleeping at that moment.

What happens when we do not care whether we are interacting with an automated avatar or a human-directed one? Is this a situation that will be healthy for us, or one that will corrupt our humanity? What if, one day, we have to limit the amount of time people are allowed to be in the virtual world by law because of the societal problems that virtual-world addictions cause? I can see this happening if we lose the ability to value what is in the physical world more than what is virtual. I can see us forgetting that there are significant benefits of having to overcome the frustrations of the physical world. For our own good we might have to force people to limit our time interacting with the virtual, much like parents have to limit the time kids play games today.

I have long predicted that the Internet itself will turn into a virtual world where we move around in a 3D space in order to find what we need. Today the Web is a 2D environment where pictures, video and text are delivered in pages. Tomorrow I believe that a Website will be more like a building that we can walk in and by maneuvering through different "rooms" we will be able to find the information or help we are looking for. There will be virtual receptionists at the front of every Website that greet visitors and ask if they can be of help. If we are going to a site that we use regularly we will drop right into the room where we need to be to complete the task we are there for. We will be able to fly over the top of the Web in order to do a search. We will be able to use gesture controls to direct our trip through the Web and will get to go where we want to go very quickly. While doing certain tasks on the Web like participating in webinars we will be able to see the other people attending in some way (most likely through their online avatars). E-commerce will be a completely different experience because we will be able to shop visually on a Webpage as if we are walking down aisles. We will touch the products we want to buy and they will be put in our shopping carts for checkout.

I am certain that the Internet itself will move to this 3D model one day. We only need the hardware (VR glasses) and tools like gesture control in order to make this happen. As soon as we have the hardware necessary we will spend years rebuilding sites to be more like offices and physical spaces instead of the pages we have now.

Virtual Spirituality

A regular theme in science fiction is the concept that people get addicted to living and working in the virtual space because they are more productive there or because they can be more successful in the virtual world than their real life. This is a very real scenario I expect to happen over the next two decades. When

I talk about dissonance this is one of the examples that will be very difficult for many people to handle: the delta between our functionality in the virtual world versus the physical world. Another challenge will be how productive we have the ability to be by using augmented reality devices versus only having to "see" true reality. If we think we are addicted to our mobile devices today, our addiction to a virtual world or to augmented reality will be much deeper and will affect us at far more fundamental levels. As wonderful as our physical world is some people will come to detest having to live in it, which could cause a problem for humanity. An alternative that some people may choose instead of suicide or drug addiction could be psychologically exiting to out on the physical world to "live" in the virtual world. There are positives to this in that it will be less physically damaging and permanent, however quitting on life is never a positive choice.

As troubling as this outcome may sound, there also is a great wellspring of hope in people wanting to live in virtual worlds for part of their days. The hope is that the virtual world becomes a stepping stone for people to open up to the understanding that our tangible world is just a physical plane, and that a higher plane exists. There is a real world that exists at the spiritual level that cannot be seen in a physical sense and we we may be able to transcend our physical world more easily if we see virtual worlds as a stepping stone.

Rudolph Steiner once wrote that, in the womb, we are totally open to receive from our mothers all that we need for our bodies. At our death we need to be totally open to receive from our Father all that we need for our souls. In Michael Mead's book, *Fate and Destiny*, he says that, "We each come to earth with a destiny that relates to our soul and fate is all that happens in our life that gives us opportunities to wake up to that destiny." God gave each of us the ability to create a world in which we can survive physically. We are meant to outgrow this world I have been calling the "real world" to recognize the world of our soul (a higher level). It may be that the greatest gift we can receive from virtuality is to recognize reality beyond the one world that we think is the real world (the physical world). We can help humanity by growing beyond the mere physical world because only then can we bring our spiritual gifts into it. This might be what will take us forward as a whole species and virtuality simply may be a step in this direction.

Most people think of a computer-generated environment as a "fake" world today. And Granted, it does not allow for the full breadth of human interaction that the real world does. As virtual worlds get more sophisticated, however, they will seem more real to us and we will gain new tools for relating in them in more human ways. As this happens we will be able to transcend space and time in more ways because we will be able to interact with recorded avatars from the past and with any avatar anywhere in the world. After a time, what we see as an avatar today will be a fully realistic replication of the real human we are interacting with. After that we will be able to interact with an avatar while the real person is controlling it in real time, or with an avatar that is automatic and just interfacing with people through programming. So you would be able to talk to me in cyberspace and it might just by my intelligent avatar standing in for me. In fact, there might be twenty of my avatars in different places meeting different people. It just depends on my ability to buy and control multiple versions of myself.

Once again the spiritual dimensions of this come into play in an interesting way because, as a human being, I have a soul. My avatar would not have a soul, though it might behave within a certain set of parameters I have chosen that represent my values and style. How confusing might the world be if we lose the ability to know with whom we really are interfacing in a virtual world? Then again we have lost that already when we are talking to someone on Twitter or Facebook and really cannot be sure who is controlling the account on the other side.

In his book, *Sonic Boom*, Gregg Easterbrook makes a point related to the fast-changing global economics that, "nearly all countries are now involved in exporting and therefore do not want to go to war with countries that import from them. This is decreasing war though that was not the conscious intent of world trade. Likewise, the WWW is creating a new world reality beyond our conscious intent. It is here that the qualities of soul are manifesting." If I call this the "real world" it is experienced without separation from (time, distance, nationalities, etc.) the world in which we live. The Golden Rule is fundamental to every religion and the Internet provides endless ways for us to be helpful to each other. We do not need to know a person because we see them physically in order to choose to be helpful. Websites like Quora.com are perfect examples of this. Quora (which stands for question or answer) provides a forum for people to ask any question and have people from the Internet Herd answer. In many cases the people answering are experts in the field, or the actual person a question has been asked about. No one is getting paid to ask or answer questions on Quora; it is simply a volunteer environment where people try to help each other out with knowledge.

We are following a holy principle when we follow the Golden Rule, be that online or offline. What is most critical with a technology-based world or environments is that we create spaces with cooperation, helpfulness, and kindness as guiding traits. While it appears that humanity has done all of the work to create the Internet, in fact we are creating a multi-layered real world that is beyond any conscious intent to move toward a more enlightened level. And if we need to go through an experimental time of virtuality to recognize that we can create a very different (and better) multi-layered world, we will be blessed by receiving the Internet and virtuality as gifts from God.

I mentioned early in the book that I coach girls' soccer. I have been coaching them for over ten years so I know some of the girls quite well. The experience I have with them when we play a game is very different than the experience of playing a game online. Both involve competition. Both involve trying to overcome a challenge by being part of a team of people. I could play online or virtual-based soccer and use the same skills and strategies that I have learned on the field for years. I could still get the thrill of scoring a goal or making a great pass to one of my online teammates. I could have the fun of playing in virtual stadiums all over the world. I could even have the benefit of not having to face the weather when I play online. However, there is much I would lose when I compare playing the game live versus virtually. When we play I always open a game by praying with the girls and sometimes one of them will say the prayer. We hold hands and bow heads and it is a quite an intimate moment before the fray. I get to see their faces as they get prepared and we have some conversations about strategy. I can sense that they are anxious and I can tell who is ready to play and who is not. When the game starts I get to observe the game and the players at levels I could not online. I get to see the players support each other on the field or help the other players up when they get knocked down. I get to see the joy when they make a good play or score a goal and the dejection when they lose. I get to see how well they support each other on the bench. I get to hear all the parents yelling supportive things to the players in the background. There are many layers of things going in in the real world that would be very hard to simulate in the virtual world. As much as the virtual world will be efficient and attractive for many reasons, it will not be able to replace the texture and richness of real life fully.

How we experience events in life can be at very different levels of meaning based on the circumstances around these moments. Being in the presence of the creator of content is very different than just consuming the content by ourselves. This dynamic has always existed. If I read the words in a book from an expert, I may get the shared wisdom into my mind. That will be a very different experience than listening to the expert share their wisdom in person with a chance for me to discuss the ideas face to face. As we grow the concept of virtuality we will increase the amount of ways we can communicate, share information, and experiences and in every case there may be some value, however, real life feeds us in a way that virtuality never can. No matter how realistic our virtual tools become, there will always be a level of energy that cannot be shared electronically that is shared in person.

My greatest hope is that our ability to create virtual worlds becomes a stepping stone towards accepting the fact that the spiritual world is also something that is very real, yet does not have a physical manifestation. If we can accept virtual worlds as being almost real to us, maybe we will be able to better accept that there are spiritual forces of connection and love that, although not visibly seen, can be felt at a deep level and connect us all.

SUMMARY:
The Very Real Impact
of Virtuality

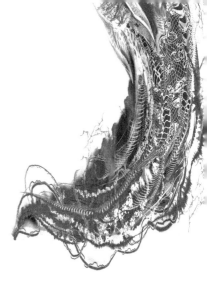

When speculating about what technology might do to us there is no larger issue than whether simulated and artificial reality will be the biggest danger facing humanity or the most productive thing we have ever invented – or both. This technology is a major crossroad for us because it allows us to create experiences and relationships that are artificial substitutes for our tangible world and, thereby, can be altered to deliver whatever we choose to be interested by. If we are in a good and healthy state, we will be attracted to simulated scenes that are uplifting and positive. If we are struggling in life, we might be attracted to a simulation that will perpetuate a downward cycle. Addictions to drugs or alcohol may pale in comparison to the future dependence on virtual worlds to help us escape from reality.

The ability to simulate a reality at will or to augment our current reality with virtual pieces unchains us from the physical reality we have lived in for all of our time on earth. Aside from our dreams while sleeping there has not been a capability to live in an alternate reality. When looking at this from a spiritual viewpoint the ability to disconnect from our physical reality will make it easier for many of us to take steps toward realizing that there is a spiritual reality that we cannot see and that is very real. Virtuality allows us to practice stepping away from physical reality; the next step could be accepting the spiritual realm as something that is very real and that will help us transcend or levels of living today. Maybe we will be able to let our souls fly free to connect with each other as if they were virtual avatars reaching across the Web to other people.

Augmenting our capabilities is the core function of technology so far in our existence, and virtualizing is an advance of a different sort. Not only is it augmenting activities we do; it is creating simulations of what is real. That creates a blurring in our minds and hearts of the distinction between what is real and what is virtual, and how we handle that blurring will determine a lot about the impact it has on humanity.

Another possibility is that virtual worlds simply are preparing us to accept that there are levels of our world today that we have not explored. There may be a spiritual plane where the methods for connecting to others is much more sophisticated than how we connect today on our physical level. We may come to understand that there is a level of energy that flows through the world that we cannot see but that is very real indeed.

I see virtuality as being very balanced in its impact on humanity so I give it a 50% positive and 50% negative score for us. I see great potential for beneficial outcomes in the business and entertainment worlds and, at the same time, dangerous paths of dissonance that will damage many people psychologically. When human beings are in pain in any way – be it physical or emotional – we tend to want to escape that current reality. Leaning too heavily on a dependence on virtual worlds to avoid the complexities of our own real worlds will have serious consequences, however.

Humalogy Viewpoint

The balance that will be critical in this area is between how much time we spend interacting in a virtual world versus in the real world. In truth, the Humalogical Balance will vary by person depending on their career, their entertainment preferences and the equipment we own. If I generalize today and factor in the amount of hours spent in virtual world environments and in simulated worlds like Second Life I still only see humanity at the level of H4.

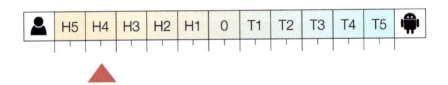

A shift to H2 would be dramatic in its implications because that would mean we are spending a substantial amount of time in virtual situations. Maybe someday we will get to a place where some people spend the majority of their work days in virtual environment. One day if we get to a 0 score where half of our time is spent in virtual environments, I will be nervous about the resulting loss of the connections in the real world that are so critical to for our health, joy, and peace.

Seminal Questions

Will technology ultimately create a virtual nirvana that entices us to to live within it instead of in the physical world? If so, how will that make us feel about our lives beyond the virtual world?

Will our skills using virtual tools ultimately create an upper and lower class based on our ability to buy the best tools and to use them actively?

Will interacting with intelligent avatars as if they are people cause us to dislike real relationships that pale in comparison?

Will the Internet turn into a 3D virtual world one day in which we can "walk around" instead of the 2D text-based world we use today?

Does God want us use virtually-created worlds and tools or is it something dangerously artificial that will wreak crippling dissonance and be the bane of our existence?

CHAPTER ELEVEN:
Humalogy 2.0 - Transhumanism?

"We are going to be a hybrid."

—Ray Kurzwell

trans·hu·man·ism (H+)

The belief or theory that the human race can evolve beyond its traditional physical and mental limitations, especially by means of science and technology.

If a word like "transhumanism" feels frightening or intimidating, just understand that it is not an outcome most people are struggling to achieve at this point. It is very likely a state we will reach whether we make a specific effort to or not. As a species we were created to evolve and, therefore, we crave progress. We seek to improve constantly on what we understand and what we can do physically. Steroid use in sports is an example of how far people are willing to go in augmenting their own bodies to gain an advantage. Some of us are very willing to modify our bodies in order to accomplish more than what is considered normal. Without doing it consciously, many of us are augmenting our capabilities by connecting with an outboard brain and adding new functionality to give us more abilities quite often. The more seamlessly we can build that outboard brain into our bodies so that we can make it even more powerful, the better. It is not transhumanism that we seek in and of itself; we want greater capability, convenience, and power that will drive us forward to transform humanity. Transhumanism will be a reality that happens at some point in our quest for progress.

Let me connect growing toward be transhuman to the Humalogy Scale before we go any further into the conceptual side of transhumanism. The scale is defining what is entirely "human" at one end and completely "machine" at the other. At the point that we find a balance in the middle and have the ability to use technology effortlessly to connect to the Universal Mind to get any piece of information or to provide any piece of content to the collective, we will have the power to be transhuman. If we were to become 75% bionic and connect so fully with the Universal Mind that we had no individuality any longer, we might be seen at a T3 or T4 on the scale and we would no longer be transhuman; we would be more cyborg than human. Being transhuman might mean that we have evolved to such a high level of capability through our technological augmentation that we cannot be categorized any longer as human alone. However, being transhuman does not need to mean that we look like the pictures of a cyborg today. It is very likely that the technology we integrate into our lives and our bodies will not be noticeable to the naked eye. It simply means we will use technology to augment ourselves to such a degree that we are vastly more powerful than people in our past. Maybe a good use for the Humalogy scale in the future will be to define the difference between being 100% original human, transhuman, cyborg and machine – from H5 to T5.

The evolution that is described in the formal definition earlier talks only about the physical (technological) and mental evolution that could grow us into a whole new species. There is no mention of our spiritual transformation though. We have had forms of religion (the worship of something as an influencer on our lives) for thousands of years. Early in our history we worshiped idols because we sought to have good luck or to improve our lives in some way. Today we consider worshiping to mythical gods like Zeus or Poseidon to be archaic or even barbaric. As the centuries rolled on we became more aware of a spiritual dimension to religion. Different religions have various views of what the spirit is and how it influences us, and I am generalizing these terms to include as many peoples' views as possible. What many believe is that there is a spiritual dimension to life given to us by God: a dimension that is not physical like a tree, computer, or human body. People who believe that there is a spiritual component to life believe that the spirit has a vast influence on us as human beings, how we act, how we connect, and how we progress. Since it is likely we are destined one day to be transhuman, will that include a big leap forward in spiritual growth also? Will we see a powerful combination of the digital and the spiritual where the Universal Mind truly brings us to a wonderful sense of Oneness?

As a follower of God I want to believe that transhumanism will include this integration of the digital and spiritual planes. I also want to believe that we will see all of our inventions support our ability to connect with each other on a worldwide scale and to do this peacefully and lovingly. In order to create a utopian world, I believe we need to set goals for what we want transhumanism to include. We need to define that word in a different way. We need for the transformation to not only be physical and intellectual; we need it to be spiritual as well.

In order to help us accept that we can and will move forward toward something called transhumanism it is helpful to look at the momentum we have achieved already since we first populated this planet. I have broken down a handful of major eras of progress that have transformed us and will transform us more in the future.

Transition one: We progressed from being cave-dwellers to learning how to communicate through language and exist in large, organized communities. Humans who could talk to each other and form large communities had great advantages by sharing work and aggregating capabilities. This allowed us to develop large scale agriculture to feed all the people who gathered together. Having a fraction of the people feed the rest allowed the others to specialize in tasks so they could advance the group standard of living through their skills.

Transition two: We made great strides in creating tools and materials (like metals, engines, and fabrication devices). We developed engineering, chemistry, biology, and made advances in the medical field. We used our new engineering and tools to construct homes, towering buildings, and even entire cities. We harnessed natural resources to generate power and delivered it to individuals. Humans with this powerful use of tools and control of resources have far outperformed those who came before them.

Transition three: We made major leaps forward by creating electronics and computers. This afforded us an amazing ability to shrink the size of our machines and to move us toward augmenting our information exchange and knowledge-storing capabilities. We could communicate at will anywhere in the world which had dramatic impacts with bringing the world together. Humans with control of these resources and capabilities have made the lives of people from the manufacturing age look crude by comparison.

Transition four: We are in the midst of this transition today, which is highlighted by integration of technology into our lives and bodies in order to enhance our performance in many ways. The difference with this transition is that the previous three were about learning to use tools and resources as external objects in better ways.

This transition will be about augmenting many of the capabilities we have natively with technology that is built into our bodies or attached to us in some way. The capabilities this will bring will make or help us behave and perform very differently than in eras-past. We will truly be "H+" (the symbol for transhumanism).

Transition five: Our heavy use of implantable connectivity will help us more quickly advance the Universal Mind. We will be connected to billions of others in real time in order to think and solve problems, and to step away and enjoy being an individual as we choose, all with minimal or no effort. Computers and software will be able to "think" in ways that mimic the human brain and will be more powerful in their abilities to process information and make decisions. The Universal mind will be more powerful than any one computer and a computer will be more powerful than any one person. This will all seem very normal to us by then and will have a tremendous impact on how we live, how long we live, and how spiritually connected we become.

Transition six: Maybe the next transition will be to transcend planet Earth. As our population grows and natural resources get stretched there may be more impetus for pioneering people to find other planets to cultivate. As they do, humanity may split into very different factions who mature in strange ways based on their environments and unique cultures. This will separate humanity into many different subspecies.

Transition seven: If we have been growing spiritually all along the way we will learn that there is a spiritual plane that is what God has meant for us to tap into all along. As we do this we will find that we do not need technology to connect us because we can connect through spiritual channels. These allow us to communicate and love each other in ways we do not fully understand and accept today. We will not be gods; we will be able to connect with each other and with God with the richness and depth always intended for us.

The concept that we might transform to be something very different than our ancestors might sound scary, yet we have been headed this way for some time. The big step this time will be the ability to change and augment our bodies with technology to give us amazing new capabilities. This is not something to be scared of; it is just something that needs to be well-thought-out before it leads to potentially tragic consequences.

The reality we need to accept and handle with care is that we have the ability to transform and transcend our human capabilities. We will begin to augment ourselves in many different ways from this point forward and that will usher in a new age for humanity. Since I believe strongly that we will do this, the bigger question for me is how far we will take it and when will it become mainstream rather than something embraced just by the digital fringe. Technology adoption is interesting because we adopt some things faster than others and it might be that we are slower to adopt technology that is invasive to our bodies. Based on the biases and beliefs we were taught as we grew up we have different tolerances for accepting changes that technology brings. We grab and run with some of our inventions the moment they are available, like online social and mobile tools. Others take a lot longer to get into the mainstream, like wearable devices and decision support systems.

Today we are making rapid progress in both biotech and nanotech, and each of these fields is moving closer to integration with information technology. As the three fields come together and advance a few steps further we will have an exponentially greater opportunity to augment our capabilities. When seen in this light it is even more obvious that we will be able to ease into transhumanism in the foreseeable future. There is no solid line to demarcate transhuman from human today. No one has laid out a set of tests that would define that we have evolved beyond our forefathers as a species. Maybe this is a test worth creating so we can recognize when we have taken this step.

At the moment we have made much more progress in improving our ability to access knowledge quickly and to communicate efficiently. We have focused physical augmentation on helping those who have been disabled. There will be a transition the moment our technology used to help someone with a physical ailment becomes so good that it surpasses normal human capabilities. At that point corporations will market their enhancements to the able-bodied because there will be lots of money to be made by helping people be better physically than they are, even if they are perfectly fit without augmentation. What people do already with plastic surgery comes to mind here. The difference will be that technology augmentation will not be about the superficial improvements; it will be about giving people a greater ability to accomplish more in their careers, have a higher quality of life, and hopefully a more developed connection with each other.

Even before we are implanting technology inside our bodies we will have increasingly sophisticated wearable devices that will provide amazing capabilities as well. Imagine an athlete who could wear special contact lenses that deliver real-time data during the game. Consider the benefit for a co-worker who uses a hearable device that allows for auditory information delivery on a constant basis or students who have highly sophisticated heads-up glasses that read, search, and deliver information on whatever is being presented visually or verbally. Either we will have to ban the use of these devices in order to have level playing fields or accept that some people will integrate technology better than others into their performances.

Some of us will fear people who are augmented. Some will be jealous of them, especially if enhanced people take away their jobs. Augmented people will have capabilities that will far surpass what a "natural" person can do, so there will be huge issues regarding fairness in hiring, pay and even popularity. We will develop all kinds of slang terms for the "Augs" who enhance themselves and, in turn, they will develop slang terms for the "Naturals." As the groups seek to make their way in the world, they will cluster with others that are like them. This will help the Augs learn how to leverage all of their new capabilities, and will allow them to develop even more ways to excel. The Naturals will isolate themselves in a non-technology world and will proclaim it to be the only true human way to live. Eventually class warfare will erupt between the Augs and the Naturals in schools and in the economy, and humanity will spend decades sorting out this problem.

Will parents lean toward augmenting their children in order to give them an edge in the world, or will they seek to keep them natural because that is the way they themselves grew up? If people stay true to the past, some parents will augment their kids in order to help them have a better life which will trigger a "keeping up with the Joneses" dynamic that will drive the enhancement of our young people. The more capabilities the augmentations give a child, the more driven some parents will be to purchase them for their babies. I hope this does not come to be, however we seem to repeat many of the mistakes from the past, and fear and bigotry have not been eradicated just yet.

The more we are prepared for this in advance, the more we can guide humanity by creating conversations, policies, social mores and, eventually, laws that will help us accept each other even as some of us migrate to an overall balance point. We have so many examples already of what can happen when we do not accept a future that is certain to come. Think about all of the times we have gotten ahead of ourselves with technology and not had a plan for handling them responsibly (nuclear fusion, texting and driving, online porn and many others). This is driven by people who cannot imagine a future that is very different than the present. They try to overlay current-day beliefs and social standards onto a future that will be very different because of the vastly different dynamics and tools we will use.

Technology has been the key to helping us tap into the potential of an extremely powerful new dimension for us: the worldwide Universal Mind that aggregates all of our individual minds to form a gigantic neural network. The network is crude today yet still allows us to share thoughts, solve problems, create new ideas, and communicate in real time. As we invent new devices and implants that allow us to connect with the grid more easily, we will connect with everyone else more easily with it too, or at least with the groups we choose to connect with. When we connect today we use the Web to share thoughts and experiences, ask questions, post information and follow what others are doing at the moment. When we can do all of this with a thought instead of clicking and typing, and when that stream of information is as close to imprinting on our minds as our own thoughts, we will have taken the next steps toward having a highly-functioning Universal Mind.

So let's be proactive in understanding how to handle this growing capability in a healthy way. How can we be part of the Universal Mind and not fall prey to group-think at the same time? How can we enjoy the powerful capabilities without also losing ourselves in an addiction to being connected to the Universal Mind? Or perhaps we were meant to be a "collective" as a species in which case this an endgame of sorts. That is an important philosophical and spiritual question to ponder. I believe we were designed to be individuals, and that we collaborate with others in order to make progress. Technology helps with this collaboration. I do not believe that we were meant to lose our individual selves in the grid and adopt group-think to guide our every belie, however. I am hopeful that the Universal mind and transhumanism are more big steps toward oneness for humanity.

I have read and heard people suggest that, once we do become transhuman, we will be god-like. The same thing has been suggested about when we form the fully-realized Universal Mind. We will have god-like capabilities and that, ultimately, we were meant to progress from being flawed humans to being gods. I guess this debate hinges heavily on what we believe a god is or who God is. I don't pretend to speak for God or to be able to delineate when we might have god-like capabilities, so please understand that I address this topic with humility. I don't believe being transhuman will make us gods. Rather, we will be augmented for a higher level of achievement. Even as we turn the Universal Mind into a powerful entity of oneness I do not believe that makes any of us individually a god no matter how augmented and connected a person becomes. What I do believe is that, the closer we get to true oneness, the closer we get to god-like qualities. Only when we transcend this world and move on to the spiritual world can we be like God with the energy we share. I am very sure that people who think we will have the power of a god because of our technology completely miss the point. When we printed bracelets with the phrase, "What Would Jesus Do?" we were on the right track. Maybe the next stage for us is a bracelet that says "What will I do with the power of God in me?"

I do believe that the spirit of God is defined by actions such as love, forgiveness, and grace for everyone (oneness). If technology and progressing toward transhumanism help us to progress in our capacity for love, forgiveness, and grace, I think this is an asset. I believe that we were meant to reflect God-like qualities such as these and that, by practicing love for each other, embodying forgiveness to a greater degree, and having grace with regard to our differences and actions, we will be like God intended us to be.

I love the thought that to claim that we have reached transhumanism, we must define this new state from both a capability and spirituality perspective. It will not mean that we are gods. It might mean that we have transcendent capabilities and that we have the spiritual maturity to know how to use them in order to make all lives better. When every human can have a healthy and progressive life we can claim then that we have transformed at a very powerful level that we can call full transhumanism in earnest. Until then, we might only achieve partial transhumanism - power without enlightenment - which could be a recipe for disaster.

SUMMARY:

Humalogy 2.0 - Transhumanism?

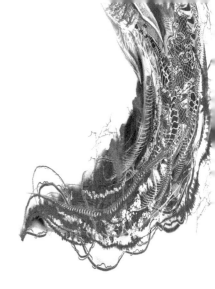

I wish fervently for humanity that we have a greater ability to look forward and extrapolate our current trajectory so that we can make better present-day decisions. We seem to be doomed in many situations to invent and discover new tools and concepts, only to deal afterward with the consequences of using these new tools instead of putting boundaries and safeguards up in advance. With nuclear science we built the capability to destroy our planet for the first time, and we have not had the maturity to assure ourselves that this is not a possible outcome. We live in danger of this happening because we invented something that is very powerful before we had a collective ability to harness the positive potential without allowing the deadly consequences to take hold.

The combination of biotech, nanotech, and information technology is going to dwarf nuclear science, both with its positive potential and its danger. This new combination will have the ability to transform our species at levels that may well make us transhuman. They also will have the ability to destroy as surely as a nuclear bomb can. There is a balance between potential and danger with technology; the greater potential, the greater the possible danger. We see this in general with the Internet today.

In order to ascend to the state of transhumanism in a healthy way we will need to combine technical and spiritual growth. This book is all about helping us understand the need to combine the spiritual with the digital. Add up all of the information in previous chapters and note that, in order to be transhuman, we will need to emphasize the God-like qualities of love, forgiveness, and grace along with brain-computer interfaces and the Universal Mind.

I put the ratio for the impact of becoming transhuman at a 60% to 40% split between our chances of ending up enlightened and in a better place versus taking steps backward as happy and well adjusted humans. The outcome of creating either a future utopia or dystopia depends on our ability to transform in balance between the digital and the spiritual worlds. We must grow as fast spiritually as we do in gaining power through technology so that our souls can guide us toward using our newfound powers in healthy and wise ways. If not we may end up with a world that looks like one depicted in *The Matrix* and *Terminator* movies. A 60/40 set of odds may not may not seem promising, but I do think God wants our world to become more enlightened and prosperous over time. That does not mean that we won't learn some hard lessons along the way about applying our technological inventions in the wrong ways. Hopefully we can avoid such consequences because we become wise enough to avoid negative outcomes. This would require a great improvement over our historical track record of employing such wisdom, however.

Seminal Questions

Is what we call transhumanism a natural next stage for our species that does not need to be demarcated?

How will we define the tipping point from being human to being transhuman?

How far will we go to integrate technology into our bodies in the quest for productivity? Body modification? Brain modification? One day will we choose to become more machine than human in a physical sense?

Will we convince ourselves that transforming to a new highly-augmented state is progress, only to find out later that we robbed ourselves of critical elements of humanity?

Will technology growth ultimately transform us to a new state where we create nirvana for our lives or will it bleed the humanity out of people who become addicted to the Universal Mind and group-think?

CHAPTER TWELVE:
Finding Harmonic Balance

If you are unaware of your thoughts, or have abdicated your choice to direct your thinking, then electronic methods can easily influence and control your mind. Your conscious mind sets the commands and the subconscious mind is the processing center where the directives or beliefs are filtered and then carried out. Empowering your life begins when you have control of your thoughts. You then must decide what you want and proceed to attain it; otherwise your commands will have no conscious intent to direct the course of your life. Happiness and peace of mind are the hallmarks of real empowerment, and they are present when you are in balance with your values and intuition.

From *Path of Empowerment* by Barbara Marciniak

While writing this book I constantly evaluated how I use technology because, after all, I am a technology strategist and digital tools are a huge part of my life. Just as I have done with the content and ideas expressed in this book, I tried to be fair-minded about the positive and negative ways I allow technology to impact my life. While writing about balance I had to admit to myself that I am often out of balance in very important areas. For this reason, I found writing the chapter extremely helpful. Every time I re-read it I committed once again to rearrange parts of my life in order to live in a better state of mind.

It is more difficult than ever to find a life-giving balance in this world. Many of us are struggling just to manage school, our careers, raising kids, or simply to survive. Investing energy to consciously build balance into our lives seems a bit of a luxury.

Yet to have a high quality of life, we must be proactive in designing a life the does not overly focus our time and attention in any one area. This includes our time spent interacting with a screen. Would you answer "yes" to any of these questions?

1. Is looking at email or a social site the last thing you do before you go to bed and the first thing you do when you wake up?

2. Do you have more conversations by volume over text messaging than in person?

3. Do you go days in a row without being outside for more than two hours straight?

4. Do you spend more time binge watching entertainment videos on your devices than reading valuable content that helps you learn and become more enlightened?

5. Do you spend more than eight hours a day in front of a screen and feel emptier inside now than ever before?

6. Has your work day extended deep into your private life because technology devices allow you to work from anywhere and at any time?

7. Has someone close to you disgustedly asked you to put down your mobile device so you could have a conversation?

If any of these resonate with you, then you might be slipping out of a good place on the Humalogical Balance. There is a good argument that anything that is unhealthy in life can be traced back to a lack of balance, be it physical, psychological, spiritual, or emotional. Too much, or not enough, of anything causes pain, and so it is with all things technological. The sinister aspect of our digital tools is they often fool us into feeling like we are connecting with people or being more productive when in truth we are missing out on critical pieces of life.

In previous chapters I used the humalogy scale to show where we are in the course of history with regard to balancing humanity in general and technology. Now I want to address finding the best balance between technology and our personal lives. We may not be conscious of this balance point today and it will get harder over time as we invent more interesting tools, yet there is a balance that will allow us to be productive, efficient and well-adjusted at the same time. Finding healthy balance in life is a constant challenge for all of us: eating in the right balance, exercising in the right balance, investing our time to balance our priorities, handling our emotions in a balanced way, and the list goes on and on. Many of us are not aware that we are far from a harmonic balance in our lives, we just get up and try to meet the challenges of each day. There is not a simple barometer that shows us at the end of each day if we equally invested our energy across the critical areas.

Consider how important balance is from a scientific standpoint. There are many elements that must be kept in balance or we see disastrous results. Our planet rotates on a gravitational path that is in a perfectly-balanced distance from the sun and it travels on ellipses that give us the seasons of the year. If this gravitational balance were disrupted by even a few percentage points life as we know it would be extinguished. Nature has all kinds of balancing factors that keep any one element from taking over and destroying others. We have learned that the extinction of any animal alters the balance of the natural world because either predators or prey can reproduce in unhealthy numbers. Even elements as small as atoms depend on the balance of the electrons and protons in order to create matter and hold it together. When we split atoms and disrupt their balance we get very large explosions. Our very lives on this planet are predicated on a fine line of balance that allows for our survival.

Men and women (the male and female energies) have many built-in traits that balance each other. In each case these energies can balance out the strengths or weaknesses of the other's energy in order to help two people create a strong personal bond. It is impossible to find peace, happiness or joy when we are out of balance in our lives; we will experience stress, danger and other negative consequences.

When I write about balance I am not talking about an equal amount of two opposite variables. Rather, I am referring to finding the proper proportions of variables in our lives so that we can be healthy over the long term. I mentioned at the beginning of the book some of the dangers of television. From a balance viewpoint violent and graphic content can jerk our balance point toward fearful and aggressive tendencies and states of mind. Sitting on the couch for hours at a time creates a sedentary lifestyle and a corresponding physical toll because it deprives us an equally important need for physical movement. Today there are far more opportunities to distract ourselves beyond television and the more technology we invent, the more possibility there is for us to sacrifice our healthy state at the altar of digital distraction.

With all of its intriguing new tools, toys, and massive amounts of information that grows at an astounding rate, technology calls to us to connect. It fills our free moments, helps solve our problems, entertains us, and stores our memories. It pulls on us to pay attention to small, medium, and large screens and away from the people around us. It directs us to a never-ending river of information and away from quiet contemplation. It pulls us toward socializing online instead of sleeping. It chimes at us regularly so that we never forget we are on the grid and have news or messages to read. Technology gives and it takes away, and has the ability to drag us so far out of balance that we pay a heavy toll.

I would like to spend most of this chapter pointing out examples of the most important areas of our lives where balance is impacted by technology. In each of these we have a choice to create a healthy state for our lives or we can spin blindly out of kilter. My hope is that, through reading about these balance points, we will be more aware of the choices we make.

Being On vs. Off the Grid

As mentioned earlier being "on the grid" is my euphemism for being connected to the Internet and available for someone to communicate with us in real time. It is the state of being digitally connected and instantly accessible. For many people I know this is not even a choice they can afford to make. They believe that being on the grid all the time is the whole point of having mobile technology and that to go off the grid is a dangerous and disrespectful thing to do to their family and friends. Others rue the day they ever bought a smart phone and became available at any moment to everybody. There are family members and bosses who have the expectation of being able to communicate with people around them instantly and any time they choose. This creates a situation where there is no ability to control the amount of outside stimulus our minds engage with. This results in little control over time for contemplative thinking.

There are times when we need to be able to relax and have confidence that we will not hear a tone or vibration signaling that someone needs our immediate attention. There are times when we need to be able to focus on something other than tweets, blogs, Facebook posts, text messages or phone calls. In order to stay healthy we need to be able to connect to other human beings in a focused way. I have observed for some years now that the people closest to me are not completely present when they have their devices in the same room. Their ability to really focus on me or what we are doing together is limited to spurts of time in between alerts on their mobile device. Invariably there will be a moment when they pick up the device and check Facebook,

Twitter, Instagram or some message on their mobile device, at which point our personal connection is lost or at least compromised.

We also need to be able to enjoy nature in a peaceful and uninterrupted way. There is something about being outside in the trees, at the beach or on a lake that centers us. Of course this has more or less impact on different people. I believe that universally we are calmed and balanced by fresh air and the energy nature gives us. These basic human needs go unfulfilled when we are on the grid and connected to technology all the time.

Being on the grid is a life-changing capability when we use it to its full potential. At the same time never being off the grid can grind a person down until they are being overly stressed on certain days. It can also create a feeling of disconnection because the digital whirlwind causes us to lose a deeper level of connection with people and the Earth which we are intended to have.

The drivers of imbalance: When we convince ourselves it is a requirement in our lives to be available every moment of every day to everyone, we have made a fateful decision. There are reasons we tell ourselves we have to be eternally connected. Examples include our belief that someone might need us instantly and we cannot accept that the world would be okay without us for a time. Or we believe that putting our device on silent so we can do something else is being off the grid (it is not because we will grab for our device the moment it vibrates). In other cases there is a deep-seated addiction to being on the grid because it fills important roles in our lives like connecting us to friends and family, to instant help or instant information. At times being on the grid represents feeling loved because people "like" what we say or post. We also become addicted to filling every spare moment with entertainment or digesting information. Maybe worst of all is the addiction to the distraction being on the grid provides because we do not want to experience our real lives. All of these are excuses being used to convince people they do not need the balancing factor of being completely off the grid for blocks of time.

The wisdom of balance: In this case the rewards of balancing being on and off the grid for the appropriate amounts of time are centered on productivity, creativity, peacefulness, and intimacy. We will be in a better state of mind if we achieve balance because we will supply ourselves with critical and needed experiences in life instead of distracting ourselves from them. Our minds need time to be in multiple states. For blocks of time we can flood our minds with new information and experiences, then we need peaceful time to digest and reflect on the fire hose of information we just consumed. In order to think deep thoughts, we often need extended periods of time without stimulation so our minds can focus on ever deepening-concepts. Investing too much time connected to the grid means that we can be over stimulated with information and communications, and far too interrupted to have deep contemplative thinking time. It is simply too much of a good thing. For these reasons it is wise to balance the states that we put our minds into.

I have backpacked in the mountains of Colorado for many years. In some ways I am happier in a tent at 10,000 feet with a campfire and no technology whatsoever than I am at my beautiful house in Oklahoma. As much as I love all my technology I also love being on a trail, deep in the wilderness and living by my wits with only a paper map as a guide. I have climbed twenty-four 14,000-foot peaks during my travels starting when I was thirteen years old. When I first started climbing these mountains we would struggle to get to the peak and then maybe run into one or two people who also made the climb. We would all take a few pictures, eat a little bit and then head back down. Then came the climb I will never forget. I was in the middle of a long backpack trip and we struggled to make it up the "fourteener" (peak that is at least 14,000 feet in elevation). As we got up to the summit we saw a young man talking on his cell phone. This was before the era of smart phones so it was probably a flip phone of some sort. He talked for about ten minutes while the rest of us listened to his conversation. I had two thoughts that day. The first was that it was amazing the cell service worked all the way up on top of the peak. My second thought was that he ruined the whole experience for the rest of us. Earlier this year I found myself checking email while I sat on a rock at 12,500 feet while doing some climbing. There used to be something mystical and challenging about climbing a peak. It pitted human strength and resolve against the elements and altitude. When the human won we got to sit at the top of the world and marvel at what we had accomplished. We were miles from most other humans and separated in a wonderful way from the crowd. This magical moment is gone for the most part because people get to the top of a mountain and call their family and friends to check in. For my part I have learned that doing email every 1,000 vertical feet is a poor balancing choice.

Stimulated vs. Contemplative

Our mind reacts to stimuli, be they externally or internally generated. In a plain white room with no people and an absence of sound, our mind can still be stimulated by any topic we choose to think about. For some of us (introverts mostly) we can survive in a plain white room for months with only our own thoughts and musings to keep us company. Others (extroverts mostly) will wither and die in such an environment. In our modern lives we almost never have a moment with such sensory deprivation; we have the opposite. We are bombarded with noises, colors, and activity flying around us at all times other than when we sleep. We fill nearly every waking moment with something we are pursuing (school work, family work or career work) or we entertain ourselves in some way, often involving some kind of screen.

The danger in this is that our minds need time and space in order to process information. Much like a computer, our minds can only process a finite amount of input in a given amount of time. Constant stimulation without an ability to do contemplative thinking results in the information and our thoughts about it vanishing as if they were a temporary vapor. Our minds can become immersed with input yet not hold onto anything that might allow us to become enlightened and to grow. Without contemplative thinking one could hold thousands of hours of input temporarily and then forget it all without contextualizing it for practical life application. This could include spending hundreds of hours of binge-watching watching Netflix and not having one enlightened thought about the meaning of life or, exchanging thousands of emails and text messages with various people and not having one new idea for improving a broken family relationship.

In all honesty I need to confess once again that I had not given a lot of thought about how I balance stimulation. I hate to be bored or waste time. This drives me to organize my days so I am very efficient and I am always making progress on something. The only time I "turn off" is when I sleep, and even then, I go into a sort of subconscious processing mode. Currently I am a recovering stimulation junkie. I force myself to step away from all my devices, from the TV and from other people for moments at a time. I am choosing to change the balance point and create even more time for contemplative thinking; I just have to get this book done first!

There is a high price to pay when being out of balance through over-stimulation and a lack of contemplative thinking. It will stunt a person's growth toward enlightenment and self-actualization, and that creates a tragic failure to reach our potential.

The drivers of imbalance: In many cases I think people are completely unaware of the amount of stimulation they receive versus time spent in contemplative thought. They are addicted to constant stimulation so they don't get bored. Like any addiction constant stimulation can seem like a positive thing at first and, like any other, it can grow more invasive in our lives. The heinous part of digital stimulation addiction is that there are few external symptoms. A drug or alcohol addict exhibits more physical or mental damage. A shopping addict has a depleted bank account. A digital stimulation addict may show no outward signs at all because the damage is one of omission. It is not what they are doing that causes the problem; it is what what they are not getting to do because of the time spent engaged with a screen.

There also are people who, for psychological reasons, run away from contemplative thinking because they would have to face behaviors or experiences they are trying hard to avoid. For them stimulation acts as a distracting painkiller.

The wisdom of balance: Our mind processes information and ideas into knowledge just like a factory processes raw materials into a finished product. We have a choice about what information and which thoughts our minds process and what the final product of that processing becomes. If we constantly process gossip, negative posts, and unimportant conversations, we will have chosen to invest our minds in worthless raw material. If, on the other hand, we choose to process valuable information that helps us learn useful things about life, we will move forward as human beings. In order to complete the effort to move forward, we must take the time to process fully, memorize, and create context for that which we have processed. This takes allowing our minds to be contemplative and relaxed.

We have control over what we learn and what we come to understand about the world and the meaning of life. There are people who are, by nature, seekers of truth. There are others who survive moment to moment and could care less about the meaning of life. Learning to set aside parts of our days consciously to do deep, contemplative thinking about what we have learned and to develop new thoughts is a critical part of maturing as a human being. This requires balancing stimulation and contemplation.

Physically Active vs. Sedentary

The human body was not constructed to be sedentary. It breaks down in a number of ways when it does not get used. There needs to be a balance between exercising and resting, and this balance is necessary at every age. This is about as basic as it gets with balancing a human need in order to stay healthy. Over the past fifty years there has been a steady decline in the physical activity people get in the U.S. There are many contributors to this including the prevalence of vehicles, the growth of television and a decline in blue-collar jobs that require physical work. The latest negative influence is all the technology we have that forces us to sit still in order to engage a small or medium-sized screen.

Whether we are sitting in a cube at work all day or playing an online game while on the couch, the impact is the same. We spend hours sitting and not moving. For those of us who have the types of jobs or hobbies that require many hours facing a screen, there must be a balancing activity that keeps us physically fit.

What happens if virtual worlds and virtual reality get so real that we can experience activities in life without really having to do the physical work to earn the experiences? What if we can see what it is like to hike to the top of a mountain without actually getting out of our chairs? What if we can play sports without ever leaving our couches? In a sense we already do that with computer games we have today. What if it becomes so easy to video chat with someone that walking 200 feet to see them in person seems like a waste? If we do not make a conscious effort to balance exercise with sitting still, we will once again pay a heavy toll.

The drivers of imbalance: Technology provides stimulation in the financial, mental, and emotional realms by giving us access to information, communication, and entertainment that can trigger the same release of chemicals in our brains. The euphoria of winning an online game can release some of the same chemicals that are released when we run three miles. Mental stimulation while sitting still does not replace the same stimulation earned by exercising, however. The touch of another person or a great workout can make us feel great at a physical level. Our brains reward certain behaviors by flooding us with chemicals that make us feel better. Being able to get technology-driven reinforcement while sitting still for hours at a time is a reward system doomed to cause physical problems. In balanced amounts, sitting still to pray or meditate can be a positive practice.

There is another driver of imbalance: exercise takes effort. It does not cost anything to go take a walk so the demand is not financial. There are some people who would rather be sedentary because it takes less effort. It is amazing to me that people can be this unthinking about life, yet there are many people too lazy to get the exercise they need in order to stay fit.

The wisdom of balance: In this case the reward for balance is a longer and more fulfilling life. As our technology gets more interesting, efficient and addictive we will have to discipline ourselves to keep our bodies healthy while also availing ourselves of all the cool features we want to use. When we are young often we believe we will be healthy and pain-free forever. Years of physical imbalance will cause muscle, joint, and circulatory issues, however. This can be avoided easily by making the conscious decision to get proper exercise every day. Stand up regularly and move around. Do yoga, take walks, run laps, ride bikes, and actually play sports (not just on a device). I really should not even have to point out the need for balance in this area because it is so obvious, yet in the U.S. we continue to struggle with health issues due to lack of activity. Don't fall into the trap of ignoring the value of physical stimulation and trade away years of life. Spending hours in front of a screen is not a good trade.

There was a moment from my younger years when I understood clearly the physical problem with screen watching. Jody was one of my best friends and we were in the living room with her parents. We had been doing a physical education test in gym class at that time and were talking about how many sit-ups we had done. Jody's parents were listening to the conversation and started kidding us about the competition we were having to see who could do the most. Jody and I were about 12 years old at the time and the fact that she did more sit-ups than me was a bit of a problem in my young mind. Being a great friend, Jody asked her parents how many they could do just to get them to butt out of the conversation. They took the challenge and both of her parents laid down on the floor to show us. I should note that her parents spent the vast majority of their time watching TV. Neither of them had a job that required physical labor in any sense so physical activity was not their thing. I will remember as long as I live when they went to do their first sit-up and could not do it. They could not, between them, even do one. They struggled and struggled but could not get into a sitting position. This made an indelible impression on me because I remember thinking at the time that I would never get into that physical state. Since that moment on I have never liked just sitting around watching TV for hours on end. I attribute my aversion to

excessive sedentary activity to my experience that day watching, two somewhat overweight people trying to do one sit-up and failing.

Retained Knowledge vs. Searchable Information

There is a logic that some people argue that goes something like this: If all knowledge is available at our fingertips through our mobile or wearable devices there is not a pressing need to learn or memorize anything. We can just reach for it instantly from the variety of devices we use and pull up whatever we seek. This logic is flawed, however, because it is not practical to lean on the Internet to provide information for every decision we make each day because we make too many of them. Moreover, without a base of learned and retained knowledge we have no foundation for creativity and innovation because we would be missing the critical raw material in the mind we need in order to create new ideas: information and context. There must be a balance between our willingness to absorb and retain knowledge and our choice to let it reside where we can get it digitally.

As searching for information gets easier, we will need to make a very conscious decision about what we choose to retain so that we can mix that knowledge into the recipe we hold locally. Look at it this way: our minds need to be our first level of information storage, the Web will be the second level, and our local devices or cloud storage will be the third. The way we use our minds in our daily lives requires us to be able to have access to certain kinds of knowledge very quickly. We can use our mind to merge many disparate pieces of information together too in order to give a situation context, solve problems, and to be creative. The Internet cannot give us information instantly just yet and it cannot replace the unique ability our minds have to integrate disparate information or to be creative.

An example of this dilemma today is the debate parents and students are having about whether memorizing facts in school makes any sense. When we give tests that review a year's worth of material both parents and students sometimes wonder why remembering every detail that was covered means anything relevant in a world where we can search instantly and recall those facts. The ultimate answer to this question is going to be that we need all the facts in our minds which we require to form the underlying base of local knowledge to accomplish what we need in life. The Internet (and our outboard brain) is not a substitute for our local brain, hence we will have to balance what we choose to store locally in our mind and what we leave archived online for fast – but not as immediate – recall when needed.

This may seem like an odd concept for those of us who have spent our lives packing as much information into our minds as possible. When viewed with our high beams illuminating the future, this begins to make more sense because it becomes clear that accessing information will be simpler with every new device we invent. The Universal Mind will be easier to tap for solutions and information. It will become an attractive thought to spare ourselves the work of memorizing pieces of data because they will be so easy to get later if we need them. We will need to choose to learn every minute of every day so that we feed our minds, or else we may become empty-headed beings who are wasting vast potential.

The drivers of imbalance: When I look toward the future I am nervous about this area of balance because many of us choose to take the easy road in life. Given a choice between working hard or doing the minimum, often we will work only as hard as needed to achieve our baseline needs or, in some cases, just to survive. As technology takes over doing more for us, including providing information instantly when we need it, many of us may become gradually more dependent on advice from the Web because it will be so much easier than thinking hard. This will feel like a wonderful convenience to many who see this as freeing our minds from daily drudgery.

The other driver of imbalance can be simple ignorance. Many of us do not think of our minds as muscles that need to be worked or as powerful assets in our lives in which we must invest. Ignorantly believing that technology can replace the roles that our own minds play is imminently dangerous to us all.

The wisdom of balance: It will be enticing to some of us to let go of the work required to memorize broad swaths of information. There are more fun things to do in life than invest many hours imprinting information locally in our minds. As children we struggle with this mightily without understanding why adults want us to absorb so many pieces of information. Hanging out with friends in person or online is a lot more fun than hours of homework. As adults often we want to learn just enough in our careers to perform the job and get our paychecks. Our new technology tools give us a lazy way out by affording us access to information and knowledge because we can just go get it at the last minute before we need it. We have to know what is proper, however, in order to take proper action.

It is critical that, going forward, we make very conscious decisions about the kinds of information we want locally in our mind and the types we can leave to be supplied from our tools on demand. This is a strange thought to many, I am sure.

In fact it is not a normal discussion in our world today. I hope that parents and educators accept that we can help the next generations tremendously by giving them a gift most of us did not have: the ability to choose what is in our local brains versus our outboard brains.

The Real vs. The Virtual

Having already read a whole chapter on things virtual, I am hoping you understand the need for balance between the tangible and the digital. It is unhealthy to live in a virtual world as a way of replacing - or hiding from - the real world. The virtual world can be fun as an entertaining suspension of reality in small doses or it can be productive for career purposes. It must be contained to these though. What it cannot become is a substitute for interacting in the real world.

When we virtualize an experience, a relationship and a location, we are creating a substitute that does not have the level of meaning or energy we would get in the real world. The danger is when we fail to realize that virtuality is not a replacement for what is real. When that happens we will drain critical areas of meaning from our lives.

We must be self-aware and then teach young people what feeds our souls with energy that nourishes us versus failing to feed our souls through addictions that have nothing to give other than momentary distraction.

The drivers of imbalance: There are many drivers of imbalance when it comes to spending too much time and energy in a virtual environment. A socially awkward person may use virtuality as a protective shield to keep from dealing with real people in real situations where they might be embarrassed. We may become addicted to the high we get while participating in activities we can do in a virtual world we cannot do in the real world, be they healthy or unhealthy. Fear can drive us to hide behind the shield of our screens and choose to connect virtually so we do not have to face an outcome we are anxious about in a real connection.

Virtuality can become a crutch to support a weakness that often will get worse because we are not not put in real-world situations where we could grow and learn. In this way the imbalance toward virtuality can be crippling to those of us who do not see it for what it is.

The wisdom of balance: When virtuality becomes a way to to stay disconnected from the real world, negative consequences will result. A hug online is nothing like a hug in the real world. A smile in a line of text is not the same as a smile on a person's face in front of us. Laughing out loud online is not as infectious as a laugh in person. The adrenaline rush from playing an online game is not the same as getting outside and getting some exercise. A relationship built completely on digital communication can only be "real" on a quasi-superficial level. The real world might be messy, frustrating and painful at times, but this is what helps us overcome challenges and to grow. An artificial world where we can escape these challenges might feel better in the short run, however it will cripple us in the long run.

On another trip to Colorado I was hiking through the Sangre de Cristo range with a friend named Dick Scar. The end of the weeklong hike was a final night spent sleeping outside with no tent on the sand dunes. It was dark and the stars were amazing without all the light pollution we normally get in a city. Dick knows the stars pretty well so as we lay in our sleeping bags looking up at the stars, he regaled us with the various constellations. After an hour or so of this, I was getting a bit tired and was starting to wish Dick knew a bit less about the stars. I made a comment to him that tomorrow we would be going back to the real world. I will never forget what he said when he looked over at me: "This is the real world; tomorrow we go back to the false world." It was one of those moments where time stopped for me. Afterward I couldn't sleep because I was so struck by what he had just said. He was right, of course; the real world is the mountains, trees, drinking right out of the creeks, fishing for dinner, and walking everywhere we went. As I lay there, now wide-awake, I realized that I had the great privilege of moving between each of those two worlds. Clearly the world we live in beyond the mountains is a real world compared to virtuality, yet his comment stuck with me in discerning what is real and why we need that in our lives.

Self-Reasoned Problem Solving vs. Online Problem Solving

Every day there are problems that arise we need to solve. Many of these issues are easy to solve and we have years of history and learned behavior to guide us to resolution. In other cases the problems we face are complicated and require information we do not have handy. Our minds can become wonderful problem-solving computers if we build creative problem-solving muscle over our lifetimes. Today there is another option to search online to see how other people would solve our problem or reach out to an online solution that is delivered electronically. This is another area to balance where we have not had to make a conscious decision before. Our only choices for solving problems prior to the technology revolution were to deal with them ourselves, avoid solving them, or get someone else to help solve them for us.

There is a skill-set that is required to be a good problem solver and there is more to it than having an awareness of tools; it also requires the will to want to solve problems personally and an ability to work patiently through various possibilities in order to find an ultimate answer. If people lose the will and patience to work through problems because they have been conditioned to have solutions provided through technology in an instant, will they then become poor problem solvers?

I am not saying that we should stop using technology to help solve problems. Problem-solving is not just about tactics; it is also about the will and the patience to find solutions. That means it is a chosen and learned behavior. It requires a balanced approach to how each of us individually builds our problem-solving muscles. If we come to believe that everything can be solved instantly and perfectly with technology only, we will become very dependent and less able to find eloquent solutions.

Much like the balance element right before this one, the skill of problem-solving is not something we are having a thoughtful discussion about. We casually observe that young people do not learn how to use the toolboxes in the garage. We notice that they are conditioned to reach for their mobile devices whenever a problem arises that cannot be solved right away. We sense that something has changed, yet we are not sure if we should worry about it. We should.

The drivers of imbalance: There are two primary causes of imbalance. The first is an over-reliance on technology because it is growing in its ability to solve problems. People may just get lazy because an online search or an application can solve so many issues these days. Slowly there becomes an expectation that problems can be resolved with a few clicks of a mobile phone and, if that cannot be done, we might not have the patience to address the problem until it gets severe.

The second is the lack of self-awareness that we have weakened our problem-solving skills. If we do not recognize that we are being crippled in some ways with technology spoiling us by making problem resolution simpler, we will raise generations of young people who cannot fully take care of themselves or resolve issues as they arise. We will fail to alter what we teach in schools and will not focus on teaching young people creative problem-solving while continuing to shove facts into their brains. And perhaps worse: we will not be aware that we are failing to give our children these valuable skills until well after the fact.

The wisdom of balance: In order to succeed at higher levels we want our next generations to be resourceful. We want them to be powerful problem-solvers because they will need to be as they will face even more problems than generations faced before them. These will be different problems from earlier times, but problems nonetheless. In order to help secure our future it is important for us to look at creative and difficult problem solving as skills that must be learned at home and in school. We need to teach them that not every problem will be solved with a Google search in under five seconds. We need to teach them that some thorny personal problems are going to take a lot of time, wise counsel, and learning in order to resolve (relationship issues, for example).

By finding balance in how we solve problems we will be able to integrate the bounty of instant information available to us with creative problem resolution in order to move forward with more velocity. By combining skills with our responsible use of tools and online instructions future generations will keep themselves from being dependent on paid help for every repair. If we reward young people for developing powerful problem-solving skills they will have the ability to progress further in life and we will all move forward collectively.

Creating vs. Copying

Creating and copying are two completely different ways of generating content. There is a huge difference between these words because one takes mental gymnastics to perform and the other does not. Taking the time to develop a quote of my own versus just looking one up on the Web both is more challenging and fulfilling. I once created this quote:

Acceptance of technology starvation is tantamount to leadership treason.

It took me many hours to craft these simple words into the exact statement and feeling I wanted to convey to a group of CEOs to whom I was speaking. This was a creative process and it was much harder than if I had looked up someone else's leadership quote and copied it. Being creative is fun and rewarding and it also is a lot more work than copying someone else's work.

Whether the task is writing an essay in school or designing a logo for our company we have to make choices about how much work we want to put into a creative task. Before the Internet it was much harder to search out someone who had done prior work and either copy it or use it as a platform to complete a task. Today an online search will bring up everything we could ever want to read or see on any subject instantly and on a scale that dwarfs a library. In some ways, we are becoming a cut-and-paste society because our operating systems give us such an easy way to capture prior work with a "ctrl-C" and paste it with a "ctrl-V."

I am not saying the Internet is lowering our creativity on the whole because there is the offsetting dynamic of "crowd-accelerated innovation" that drives the fast proliferation of ideas. But individually the ability to copy the work of others lowers the inspiration for many people to be creative on their own.

By allowing technology to help us refer to prior work so simply instead of creating something new we could slow the progress innovation would have brought to us all.

There can be artistry in doing mashups of prior work and we have seen this in music, art, and movies. There are people who cannot create something new from scratch who are very adept at pulling disparate pieces together into something that is new as a whole. In this case the Web can be very helpful because it gives us access to such a huge base of raw material to mix together. However, if everyone becomes a mashup artist then we might run out of new combinations for the base source of material at some point. We need people who can create source material or solutions so we continue to progress. If the music world did not continue to evolve with new styles of music we would have the same one thousand songs to listen to the rest of our lives.

The drivers of imbalance: The Internet has such a powerful ability to put anything that has ever been done before at our fingertips. It is so easy when writing, designing, solving problems, or building to search for what has been done already and to copy or aggregate. If a task can be shortened or completed more easily by using existing art there is a powerful attraction for many to take this easy way out. There are many people who, given a choice between the easy way or the harder way, will choose the easy way. It is easier to copy someone else's work than to be creative and do our own.

If you need a specific example of how this happens look, no further than how young people do homework today. In decades past we had to read in order to absorb ideas and then write a paper by hand to consolidate and restate the topic in our own words. If we copied the text right out of a book it was pretty easy for the teacher to tell and we would be punished for plagiarizing another's work. Today a large amount of "writing" that gets done by students is simply cutting and pasting, and then editing just enough not to get caught by the online tools teachers use to verify originality.

The wisdom of balance: At the individual level if we do not make conscious choices to create something unique regularly we will not fully bring our individual God-given gifts to the world. By being willing to take the harder road and create something new that has never existed we move the world forward, even if just a tiny amount. We feel more fulfilled when we know we are adding something valuable to the world's collection.

Collectively if an increasing number of us choose to not exercise our creativity, our progress as a species will slow considerably. What if the skill we have always had to invent and create slows dramatically as more people repackage what has already been done?

I believe that we all have God-given creative abilities and talents. We have to choose to use them which requires a balance between being willing to create something new from the ground, up and leveraging prior work from someone else to complete the task. Creating is harder than copying. Creating is more noble than copying. Creating is a key component in advancing anything, including humanity.

In-Person vs. Online Communication

There is an important difference in context and understanding between being with a person face-to-face and communicating with them over the wire. There is an energy that is shared between us when we are close to each other physically that is not exchanged electronically. There also are visual clues that can be read in person that do not come across in the written word. If video conferencing is used these can be picked up. Still, video conferencing does not provide the energy exchange that a direct conversation provides. For this reason we need to make balanced decisions about when we talk in person and when we do it online. I mean this at the specific level of knowing when it is time to talk to a person who is two feet away and in the aggregate of balancing how much we use social media, email, and texting versus being present with people directly.

At times we use technology in order to have digital courage to say things to people over the wire we would never say in person. We use digital communication to hide from saying things we think the other person might not take well. Sometimes we use online communication to allow us to hold a fantasy about the other person because if we talked to them face-to-face we know subconsciously they would not be everything we wish they were. None of these situations is healthy for the relationship with the person we are communicating with.

So the question for all of us is: at what point will we start making more thoughtful decisions about how we choose to communicate? When will the pendulum swing back toward valuing a more direct conversation over one that is dashed off on our devices? As an introvert I tend to lean toward using technology to "talk" to people because it is efficient and I don't have to deal with the elements of people that exhaust me. With that said I make a conscious effort to talk to people face-to-face or even on the phone when I know it is the right thing to do. Will we have to teach courses in schools someday on how to communicate appropriately so that young people learn how to balance time communicating through screen with time talking face-to-face?

The drivers of imbalance: It is so easy and efficient to communicate with many people through digital tools. I get over one hundred emails a day which allows me to have many more conversations than I could have had years ago. I text people on and off all day, allowing me to connect with them instantly to send a picture, thought, or question. I can post updates online to thousands of people who know me and provide at least a quick connection with all of them without having to take the time to connect with any of them.

Online tools give us the ability to put emotional distance between ourselves and a person we are delivering information to who may not like what we have to say. Talking to people digitally affords us the ability to say things we would be way too embarrassed to say in person. Finally, connecting over the wire fulfills the desire to create a fantasy relationship with a person because we don't have to face what they are like in reality. Our need for emotional fulfillment can be a powerful driver of behavior and online tools can make it very easy to find someone somewhere in the world to give us what we desire. All of these put us on a path toward either unhealthy communications and relationships or vapid, hollow relationships that leave us feeling empty.

The wisdom of balance: We must develop more of a sense of balance between communicating conveniently versus effectively and meaningfully. An email providing moral support is not a bad thing to do, of course. A face-to-face conversation that shows empathy and is finished with a hug is a completely different way to provide moral support though. A text message to grandma is not the same as stopping by for a few minutes to say hello in person. In order to relate to each other as caring people we must be willing to balance an in-person conversation that may be less convenient with the ease of sending a digital message.

There is a wonderful place for digital communication in our lives but it cannot become the basis for the bulk of our relational interaction. If we trade the quantity of conversations through technology for the quality of conversation directly with a person we will find that life becomes a bit emptier. It is impossible to learn about the whole person when communicating electronically which also means that people cannot learn about all that is "us." To know and to be known is a strong need for humans, and digital communication can only provide part of the story. The most important parts of our lives are based the relationships we have with others so it is no small thing when we get out of balance in how we communicate!

Individualism vs. The Universal Mind

We took the first step toward creating a universal mind when we invented the Internet. We have been making greater advances ever since. Even though we are early in the maturation of the Web we have harnessed the ability to form a "crowd" of users already that can benefit the whole. However, we were built to be individuals too and there are benefits to having both the access to the Universal Mind and to our own independent thinking.

We are storing and delivering our created and collected information instantly for any member of the universal mind. We have built social tools that allow any member of the hive to communicate instantly with any or all, members. We can share ratings on products and services to help each know what is high quality. We can provide instant answers for each other for problems that come up. We can create instant insight into traffic problems by using an application geared for this purpose. We created crowdsourcing sites to connect people to get work done and Q&A sites where the Internet herd can answer any question someone asks. Actually we are closer to having a real time universal mind than most people realize.

There are dangers in this progressive step, not the least of which is "group think." The easier it is to see how everyone else thinks, the easier it is to accept whatever opinion or viewpoint the crowd has instead of forming our own. I believe that we were designed to be both individuals and to connect in groups. Too much of one or the other destroys a carefully crafted balance in life. For humanity to continue to become more civilized and enlightened we must balance the strengths of aggregating our collective knowledge and skills while also valuing the greatness of individual thought, debate, and perspective.

The drivers of imbalance: Many people have a strong desire to fit in, to be accepted as part of a group. In the quest to feel this way they will adopt whatever the group says is the truth. This crowd-following has led us to tragic cult-like activities and radical groups doing heinous and vicious acts. With the reach and immersive nature of online technologies there is the potential that the "group" people might want to fit in with is billions in size and is online. Another risk is that the truth a person seeks is delivered through an online search without any independent thought about whether it is the Truth with a capital T. It can be hard to think differently than the people around us if we see that that group is millions strong. It can be scary to stand by ourselves and promote the Truth when others want to pull us into the herd. If we submit to that pressure, we will be driven to lose the needed balance between having a unique viewpoint and joining the universal mind.

The wisdom of balance: We were meant to be unique. We were not designed to be exact replicas of each other or be exactly like any other specific person. There are no two people who are exactly alike physically; even identical twins are not technically identical. When it comes to the complexity of who we are there are no two people who have ever been even close to identical. We all develop completely unique views of the world, moral systems, spiritual beliefs, relationships, and life experiences. It is through this uniqueness that our species progresses over time.

It is that uniqueness which provides necessary value when aggregated into a collective that can help each of us with real-time efficiency. There are wonderful assets of teamwork and there are strengths in being individualistic as well. Teams can solve larger problems faster than individuals can. Collectively, teams can hold a much larger body of knowledge. Teams can divide skills and knowledge sets so that the whole is much bigger than the parts. Individuals can think independently so that entirely new ways of accomplishing solutions can be brainstormed. In order for the world to be a balanced place and mature at a faster rate, each of us will need to be conscious of the time we spend leaning on the Universal Mind for work, answers and ideas versus doing our own independent thinking. We will also need to be very careful in adopting what the herd says is the right answer to a question without applying some of our own independent thinking.

Humalogy in Balance

If we were to make progress with each of these examples let's take a look at what an overall balanced life might look like. A warning first: if you are a person who refuses to go a minute of the day without checking your mobile device this description might come as a shock. You will have to put it down and walk away at times. I know for safety reasons you might have to keep it close to you so we can be reasonable about this as long as you realize that disconnecting means releasing the opportunity for people to interrupt you constantly.

Let's take a typical day from the very beginning and paint a picture of the habits that will help us be more balanced.

When we wake up we engage with a human being for at least as much time as we engage with our computer or mobile device. This is easier if we are married and have children because we have someone close to us that we can interrelate with right away. It is not easy when we live alone or are traveling alone. In this case it is important not to go right from an early hour of interacting with a mobile device to the start of a work or school day where we will be interfacing with a laptop or tablet. A conscious attempt must be made to talk to someone face-to-face for awhile. For many of us finding the balance between interacting with a screen and a person in the first hour of our day is already a problem!

As we go about our day we try to learn something new. Every day is a chance to put new information into our heads so we have it stored and handy. We take some pleasure in storing information locally so that we can pull it out without reaching for a device. This needs to be a conscious decision for many people. There is an unconscious pull to be active and get things done without a specific effort to learn more. Without vocalizing it, many of us know that most answers are a Google search away so we do not feel the same call to read and learn as was necessary in the past when providing information was more likely stored in our own brains. We need to defy this imbalanced approach and seek to turn lots of information into knowledge and get it archived in our own minds for use later.

When faced with a problem we must make a call about whether this is something that we have time and energy to solve without searching for someone else online to do it for us. It is okay to reach out for advice online. It is also fine to get instructions online to fix something that we do not know how to repair on our own. Just this minute my iPhone died – went totally stone dead – and I know the battery was not out. I went online and searched the message boards and found a post where someone said to toggle the mute button back and forth and then try to reboot it, and it worked! I have watched my son do repairs on his car that he would never be able to do without online instructions. The point is that we must exercise our problem-solving muscles and make an effort to figure things out even if we do not have online instructions.

While out and about, at work, or even at home we need to talk to people face-to-face. We must make a point of it. Look for moments to take advantage of this form of connection. Look them in the eyes and communicate with them through this rich and multilayered method of connecting. Don't take any conversation for granted; each one is a chance to lift someone up, be lifted up, learn something, or help teach. We might be introverts and may dislike talking to strangers but we need to make the effort anyway because it gets easier over time and is a better alternative than being recluses. We should make sure we spend at least as much time talking to people in person as we do with our mobile devices. If we spend four hours a day texting and only one hour talking to another human, we might be out of balance.

While we are doing all of this we need to get up and walk around. We should make a specific effort to get those 10,000 steps in each day. There are many ways to make sure we are moving and exercising our bodies. We do not have to be Olympic athletes in order to be healthy. We do need to move our bodies and not spend hours lying around watching Netflix, texting people, and looking at Twitter, Facebook or Instagram. Exercising is a good chance to put down our technology and walk away for a while. I have noticed that when I run lately I don't take my phone or listen to music on it. This stopped the calls coming in while I am running and gave me some peace and quiet for a while.

If the weather allows we need to get outside and breathe some fresh air. Too many of us have indoor jobs where the only outdoors we get are about 500 feet walking across a parking lot. We were built to be outside in nature. We were not built to be inside buildings for days on end. Think about that for a minute. Our bodies were not constructed to sit in chairs or lie on beds all day. We were not mechanically constructed to do that. We were built to be ambulatory, to breathe clean air, and to use our arms for something more than moving our hands across keyboards. We were designed to do physical work, to hunt, to grow food, and to dance. If we are always indoors because of the way we have structured our lives and our careers, we must make a concerted effort to get outside for extended blocks of time.

321

When we are relaxing at night we should find another person to talk with face-to-face. If we spend four hours solid before bedtime watching TV, playing online games, doing social media, or working on laptops and fail to engage with someone in a meaningful way we are making a trade that might seem fun but that will throw us out of balance. This needs to be a conscious effort to calculate the number of hours engaged with a device and balance those with time engaged with a person or people. When I say balanced, I do mean 50/50 and not 90% living on a screen and 10% with another person.

Before we fall asleep we must not have our mobile device be the last thing we look at. We can lie there and review our day in our mind's eye if we are alone or we can talk to someone if they are close by. We must quiet our mind in order to send us into dreamland. This can only happen if we leave our outboard brain at least a couple feet away. It is important for us to turn it to silent so it is not buzzing and ringing all night, unless our jobs absolutely require it. Too many of us are interacting with a screen of some sort until we collapse in exhaustion into sleep. The problem with this is that screens have shown to delay the amount of sleep we get. The end of the day is a great chance to balance contemplative thinking with device usage by putting the screen down and just consolidate our thoughts before calling it a day and falling asleep

We must consider the model we provide for the young people around us. They learn about balance from what they see in their elders. Every bad technology habit we have will likely be passed on to them. If we have kids who live out of balance with technology - and many do - it is even more incumbent on us to be a force for good and to help them learn the value of balance. It is easier to say nothing and let young people do what they want with their devices but if we do we are not doing our jobs and we are letting our kids down.

SUMMARY:
Finding Harmonic Balance

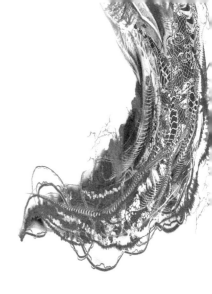

This chapter is prescriptive, meaning that it provides a critical answer to the question: "How can we make sure that technology has a positive impact and not a negative one?" Finding balance in life will not help us avoid every problem that could come our way but it will lower the odds that we will make unconscious decisions throughout our lives that come back to haunt us later. Overusing any one tool is not a good thing when life calls for the use of many different tools. Our minds and our souls are the two most important tools we have to help us through life. If we starve either of them because we think technology will replace the role they play in our lives, we will never become more enlightened than we are today.

Let's use technology to augment our minds and to help feed our souls in appropriate ways, and not over-use the digital until we are restricted from growing and spiritually crippled by it. Technology has the capacity to free us and teach us how to grow spiritually. As digital tools help us connect with each other in powerful ways either we can use that capability to feed our souls or to starve them. The ultimate cost of imbalance will be that we are alive in a worldly sense and dead in a spiritual sense. Millions of people will learn this late in their lives as they look back and see the results of the decisions they made and the consequences they ended up with. I pray for us that, by making a conscious effort today to balance what feeds our souls with what feeds our mobile devices, we will find love, joy, and peace. If this book accomplishes only that, I will be overjoyed.

Humalogy Viewpoint

There is no way to generalize where we are as humanity because there is too much variation among all of us. What I would like for each of us to do is look at the graphic below and ponder for a moment where we think we are in our lives. When it comes to the hours in our day, how many are spent interacting with people versus interacting with a screen?

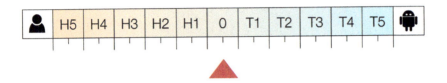

I have a strong feeling that, one day long into the future, we will come to understand that a score of zero on this scale – an equal balance between using technology and using our human skills and talents – will be the healthy and optimum place to be when aggregated across all things in our lives. When I read science fiction stories about a person's brain being uploaded into cyberspace so their consciousness can "live" without a body I always wonder how long that consciousness would stay happy and sane. I can see how moving too far toward the T side could decimate the humanity in a person. On the other hand, I can see how swearing off any use of technology in a modern world could make it difficult to be productive and to prosper. I do believe that if one had to make a choice either to be closer to the T side or H side and still remain in a good, healthy state, leaning more toward the H side is the safer place to be. We can live without an ever increasing amount of technology – it is possible to scale back a bit. Yes, it might hurt our careers in some ways, but we have been making this kind of trade off forever as we seek to have a work/life balance. There is an undeniable toll we pay if we lean too far into the T side at the expense of our human needs.

Seminal Questions

How dire might the consequences get for each one of us before we learn about the critical importance of balance?

Where are you out of balance in your life today with technology?

Are you willing to experiment with investing your time differently in order to have more of a balance?

What kind of model for keeping healthy balances are you for the people around you?

How can each of us, by finding healthy balances in our own lives, change the global direction of humanity away from a focus on the physical things to the more soul-inspired parts of life?

HUMALOGICAL CHOICES

God loved the birds and invented trees. Man loved the birds and invented cages.

—Jacques Deval

We choose what we create and how we use those creations. The invention of the Internet will be seen historically as one of our finest moments. Now we get to choose how we integrate it into our lives. We have choices about how we raise our kids and what we teach them is healthy and unhealthy about integrating digital tools and uses into their lives. We choose many times each day. Whether we make conscious choices, healthy choices or poor choices is up to us entirely. We have free will and with that comes the reality that our technology choices have consequences, be they empowering or destructive.

The ability to be a high beam thinker when making these choices is very important so that our decisions are wise. When a leader is visionary while running a business they can make wise investments that will keep the entity relevant in the future. The issue a business can run into is the market changing faster than the entity can plan for and, over time, their products, processes, and skills all can become dated and no longer valuable. On the larger scale of the human race our ability to extrapolate trends and be high-beam with what technology will do for us can give us the ability to make wise choices about our future. If a business makes a mistake with where the future might go they go bankrupt. If humanity makes a mistake with how technology will impact us it could cripple us spiritually and destroy us physically.

Stephen Hawking, one of the world's most expansive thinkers, has been taking on a new subject lately. Instead of teaching us about space and physics he is warning us about the dangers of Artificial Intelligences (AIs). When he looks forward with his high beams on, he sees that software that is self-learning and also self-directed could make decisions about taking actions that the AI believes are right and would be devastating to human beings. Consider self-driving software controlling our car that sees a child in the road with no time to avoid the child without rolling the car. The car notes instantly that there are five people in the car and only one child in the road so the car chooses to hit and kill the child to "save" the people in the car. This is a very real example and it demonstrates that, as we give more control to software, it might make decisions we would not make in that instant. Whatever decisions we program and AI to do it will do and we will have to live with those consequences.

Stephen Hawking is not alone in concern for self-guided technology. There are groups of people today who are working on rules they hope every country will adopt concerning the development of self-guided drones and robots, especially if they are of the military variety. If we take the step of allowing physical devices to make their own decisions through smart software without being human-guided, we create the potential for machines to run amok. If you think this is too "science fiction" for you, consider that we are doing exactly this when we convert from human-controlled vehicles to self-driving ones, and we are headed that way right now. If we extend this kind of technology blindly to self-guided military drones that are flying around the world hunting down specific people, and to self-guided ships that patrol the seas with no human being in control, we can start to see the dangers. Will we seek to build the ability to allow every machine to be self-guided? Of course we will, and we are a long way down that road already. If a machine can work on its own, we save the costs or dangers of a human controlling the machine. What we have not done is give enough thought to the rules and boundaries we want to put around the growing use of self-guidance by machines and software.

I have never liked the umbrella term "robots." This is a word that tries to sum up too much in one concept. Surely we will have machines that are human-controlled, machines that are self-controlled, and machines that are hybrids. Today we use the word "robot" for any of the above configurations. I suspect we will need to differentiate among these three categories and create rules for all three. So if a machine is controlling something that has mass-destruction capabilities (a plane, a refinery, nuclear plant, a battleship, a tank, or a dam), should a human at least control it so there is some discretion involved? Or is it the opposite? Do we not want humans involved in operating high-risk machines because we cannot trust everyone? I use this single glimpse into the future as one small example why it is critical for us to look forward and extrapolate as accurately as we can what technology might do if it is not programmed perfectly. Surely it makes sense to be proactive with how we develop and secure self-guided machines and AIs so they do not harm us. Will we have the foresight - the high beams - to put safeguards in place before we learn the hard way why they are needed?

The clearer our future picture is, the easier it will be to make wise choices today. If we knew for sure that, by allowing a child to integrate with technology in inappropriate ways, we would be harming them greatly as an adult we would never allow it. We are making many decisions that relate to technology use every day and, in many cases, giving little thought about the long-term costs. Seeing into the future accurately and willingly gives us the ability to make wiser decisions today. That is what I want for all of us.

I am hopeful that, through the pictures of the future I have painted in this book, you will be able to have a clearer understanding of how far-reaching the impact of our march toward a Humalogical world will be. If we can see these together as a species, we can be wiser about our responses to our progress with new tools and their impacts. I am clear that any tool we invent can be used to uplift or to destroy, at our choosing.

The Internet is allowing us to flow tremendous amounts of information across the globe with a click of a button. We can store a massive amount of our accumulated knowledge and record our collective history in digital formats for analysis and learning in the future. That means, hundreds of years from now, historians will be able to access mountains of digital jetsam and flotsam from our time in ways no present-day historian can come close to doing. Biotech and nanotech sciences are allowing us to change our world at the cellular level which is a whole new frontier of technology that also will have dramatic influences on us. Until then the Internet is the technological platform that is driving an era of progress with historical significance.

Did God create the Internet?

I do not dare speak for God, so I am just making my best effort to reason out how an all-powerful creator might feel about where we are going with all things digital. Because I believe we are becoming more enlightened slowly as a species, it makes sense to me that the Internet is a purposeful step. The Internet is a web of connection that mimics physically what God seems to want spiritually. Maybe the Web is a test for us to see how well we handle a higher level of connection (the Universal Mind) than we have had before. Certainly the Web is a digital layer of infrastructure that amplifies the Web of humanity that has existed through verbal conversations and relationships for millennia. Where once our connections were few, separated by geography and not enjoyed in real time, today there is very little separation or friction among billions of people. Once we conquer the Web layer of connection we might be led to take the next step and create human connections that are worldwide and real-time without the need for the Internet infrastructure. Our connections will be made at a spiritual level that few of us are aware of today. At that point we will graduate from being a Universal Mind to full Oneness.

I am old enough to have had a long time to wrestle with trying to figure out how to "feel better" when I am not in a good place. I graduated from thinking that drinking a little alcohol was an easy solution to getting myself out of a funk. I also transcended from thinking that some kind of entertainment could make a lasting dent in my mild depression. I have spent thousands of hours evaluating the causes of what might send me into being in a dark place and then trying to fix those causes. After years of chasing some magic fix to make myself feel better when I was low, finally I realized what could help me move to a place of peace and joy instantly. The answer is simple and elegant and it works for me every time I need to climb out of a valley. I remember the first time I tried it. I was sitting in my office at home and I just felt completely at the end of my rope. I was bouncing off the bottom emotionally. In that moment I heard a voice tell me to reach out to three people and show them some kindness. Immediately I emailed three people I knew and expressed kind words about how much I cared about them and admired what they were doing in life. I did not expect anything in return. The transaction ended the moment I hit the send button. That act made me feel better immediately. In that moment that day I learned that connecting with people lovingly and with gratitude – even over the Web – was the "magic" move that nudged me into a better place. When I get to a dark place I also feel very alone. By reaching out to be a blessing to another person I understand immediately that I am not alone and

that I can make a positive difference in the world no matter what state I might be in at the moment. I pray for you that you see we are meant to have Oneness with other people and that connecting with love is the most potent way to find peace. As I write this it is my fondest desire that you can feel a moment of connection with me through this book.

A creator set this universe in motion and somewhere gave us a desire to connect soul-to-soul along the way in order to flow the energy of love around our world. There are two very basic connections in our lives: the first is to the spirit of God and the second is to each other. We are each on unique journeys through our lives and our personal fulfillment will not be about reaching a commercial or status level. We will find true fulfillment through soul-to-soul connections we create throughout our lives. True success will be finding love, joy, and peace, and these will come from maturing our souls' connections to each other and to a holy spirit. Either we will enrich the world having been here or we will die having been consumers, leaving a deficit to show for our time in this realm.

We have been growing smarter as a species since the day we could have our first thoughts. We have been inventing new tools since we hit this world. For thousands of years God has sanctioned our need to grow intellectually. If God did not give us the free will to choose a higher level of connection we would have been stuck as ignorant animals. We would not have been allowed to have such a sophisticated ability to add knowledge or been given the choice to make our own decisions. We have made extraordinary advances in understanding the world around us, why it works and how to correct choices for better results. The pace of our abilities to invent ways to alter the world is speeding up. This results in the potential for us to advance spiritually at a faster rate as well. I don't believe this is all random chance so there must be an endgame to the path we are on. That endgame could be transforming our world over time into a much more holy place. At the same time there is the potential to misuse all this power we are being given which could set back the human race by decades as we learn painful lessons.

We were created first to be children and then to mature into adults by adding wisdom along the way. With each new generation there is progress as the wisdom from one generation was passed to the next. At first this was done verbally and with no other options. Then we created written language and wisdom could be passed along through books.

Today, and from now on, we have a digital archive that is the most powerful tool we have ever developed to retain and pass on wisdom from one generation to the next. We have grown from intellectual children into intellectual adults over thousands of years. The next step forward for us may very well be to mature from being spiritual children into spiritual beings who do not spend so much of our lives connected to the physical things of this world.

Another part of the plan may be to amplify what is uniquely human in each of us. If we are made in God's image then it is reasonable to assume that there are dimensions of our humanness that are wonderful and godly. Our existence to this point has been stunted by doing mechanical tasks for survival that have held us back from connecting with the spirit inside of us. By having technology take over more of what we do today that can be done by machines and software we are led to find what is uniquely human about ourselves and to expand on those skills and talents. What we might find is that there potential in our power that we have not even touched on yet. We might be guided to discover what is uniquely human that technology cannot do. Once we are clear about that we can expand our capabilities in these areas. We will stop having people spend all day on an assembly line, in a factory, putting a nut on a bolt, shuffling paper files and we will free them to be creative, connected, and to flow their energy to other people. Each soul comes with unique gifts and we will have more time to share these with each other.

If this is true we are in for a very interesting ride over the next century as we discover a wealth of capabilities we simply do not understand or use today. Will we discover that we have mental telepathy with each other? Will we discover that we can alter the outcome of things through creating a positive level of love and energy? Maybe we will be able to eradicate war, crime, and disease through applying a combination of our humanness and the spirit inside of us. Maybe we can make this world more of a nirvana by connecting with the spirit inside of us and empowering our growth as a species. Maybe the growth will not be just through empowering devices, maybe it will be our own internal empowerment. This sounds like a future that is very inviting to me!

In order to improve the odds we have of realizing this future we must be realistic about the negative aspects of technology I have mentioned throughout this book. I choose to believe that we will not have a worldwide apocalyptic event that will send us spiraling backward. I do believe there will continue to be pockets of the world that endure wars, bigotry, and terrorism. There will be uneven progress across the globe. I see the human race continuing to progress toward transhumanism and, as I mentioned, the evolution toward this new state will not be just about technological change. It will also be a heart change in people as they align more toward love and connecting with each other instead of fighting and killing each other. I don't say this because I am a pacifist. I say this because I have faith in humanity to "grow out" of thinking that violence and hate ultimately are acceptable for any reason. Over the centuries we have proven that we are becoming more humane (albeit gradually) and I see no reason that technology will do anything but amplify this.

Since I think we can agree that the human race has become slowly – even painfully so – more enlightened over time, technology has the ability to amplify and speed up the direction we are headed already. That is, after all, what technology does very well; it amplifies our capabilities. Rarely does it do anything completely new. Rather it scales up our ability to do something more efficiently that we have been able to do already. Over time the more civilized cultures in the world have been adhering to universal principles such as The Golden Rule, equality of people, caring for the environment, justice for all, improving education and helping the less fortunate. Technology seems to be helping us actively support these principles.

We have the collective ability to make the Internet a healthy asset for mankind. This amazing technological ability to connect us all and to store our aggregated knowledge likely will go down as one of the most influential tools we will ever invent. We can make it a safe space, a sacred space. Thousands of people have cooperated to create the Web, many as a service with no intention of monetary gain, which is a great statement on the altruism of these many folks. They dedicated their time to create and help maintain this capability. Each of us will choose individually how we use the Web, when we are connected to it, what we add to it, and what we will put up with from it. Multiply this times billions of people and we will dictate what kind of impact this will have on succeeding generations.

I believe that the Internet has more of a positive impact on humanity than a negative one. I do believe that God helped us to form the Internet and that it is a digital manifestation of many holy principles. I don't believe that we understand entirely how powerful it will be in its impact over time. Today it is a beneficial tool for most people. One day we will look back and understand what was positive and the good that came out of it, and what was injurious to us and the consequences many will have paid.

Humanity has been made up of billions of individuals who group by families, geography, nationality, gender, age, and other distinguishing criteria. As we transform toward a Universal Mind it is my greatest hope that we cease to separate ourselves in ways that cause tragic pain and outcomes. Transformation and transcendence happens one person at a time. Making wise decisions with technology starts with you and me. Thank you for reading this book; I hope it has created a new perspective for you so that you can be a glowing light in the world.

My Hopes for a Balanced Humalogical Future

1. The integration of technology and humanity steadily improves our quality of life.

2. Technology continues to free time for us through automating processes so that we can reinvest our time in tasks that require uniquely human skills, and maybe we can slow down our pace of life a bit.

3. We will be freed to have more fun, have intriguing new methods for having fun and have healthier choices for having fun.

4. We will continue to make progress toward creating the Universal Mind (Oneness) through advanced online connections.

5. Humalogy continues to improve our abilities to solve problems, be they large or small.

6. Humalogy continues to break down geographic barriers so we are truly one world and one people.

7. Humalogy helps us gain wisdom faster so that we increase the pace toward being enlightened.

8. Humalogy helps us to be at peace personally and politically.

9. Humalogy frees us to connect spiritually at levels well above where we exist today.

10. God will use technological advancements to lead us toward a more utopian future and not one where machines take over our lives.

My big bonus hope!!! One day I will have an online avatar that can do half of my work for me and that this happens before I retire so I actually get the benefit. With my luck it will be released to the public two weeks after I stop working.

"The Lord says: These people come near to me with their mouth and honor me with their lips, but their hearts are far from me. Their worship of me is based on merely human rules they have been taught."

Isaiah 29:13 NIV

GLOSSARY

Note to the Reader: The definitions I have provided are not copied from a dictionary. These are my layperson explanations of these words so that it will be easier for you understand what they mean and the context in which I am using them.

Android – So as not to confuse things, when I use this word I am referring to a human looking and acting robot. An android has no human parts which differentiates it from a cyborg. I am NOT using this term to talk about the Google operating system of the same name.

Artificial Intelligence (AI) – When software has the ability to process information at a level that mimics how the human brain might process it is said to have achieved artificial intelligence. There is not a specific test that defines whether software is an AI at this time, the moniker is normally self labeled by the companies developing the application.

Augmented Reality – This capability is best described as a method by which information is added to the viewpoint of a user. Using some kind of device, goggles, or a mobile device, the user points at an object or scene and the device adds information to the view. This can be added as text or a picture into the view of the users.

Avatar – The context by which I am using this term in the book is a software based graphical representation of a person. I can create an avatar to represent me in conversations online that looks just like me, or has any physical characteristics I choose. The avatar can be simply a graphic in some situations, or could have intelligent software that creates a personality and capabilities to act independently on screen in other cases.

Badvocacy – When a person speaks poorly of another person or company online it is described as badvocacy. This is a play on the opposite dynamic which is advocacy.

Big Data – This is a sweeping term that generally describes our capability to be able to gather, aggregate, and store huge amounts of data. With this capability, organizations are better able to improve their ability to see trends and anomalies in the larger sets of data. In many cases more data is better and harvesting large amounts is referred to as Big Data.

Blogger/Blogging – The word comes from the shortening of the original term which was "weblog." It describes the process of writing a page or so of content and posting it online for a specific group or the entire Web. Blogging allows any person with an Internet connection to publish their thoughts to everyone else on the Internet.

Bluetooth – This is the name of a specific wireless connection method between two devices. This standard is often used as a connection between mobile devices and other technologies. It has now branched out to connections in vehicles, medical devices and wearable devices.

Brain Computer Interface (BCI) – We have developed the capability to physically connect a person's brain to technology devices so that the user can control those devices just by thinking. It is used primarily in prosthetics today as an internal BCI (below the scalp). There are commercially available versions that can be worn on the outside of a users head that can be trained to do rudimentary tasks on a computer.

Business Intelligence – This is a broad term that relates to mining the value from data. It incudes many processes like data visualization and analytics, and also the tools that provide configuring raw data so that it is more useful and valuable.

Catfishing – When a person creates a fake profile online and then proceeds to deceive another user by building a relationship with that fake profile, they are described as "catfishing." This normally refers to social relationships online and not using fake profiles to deceive someone to do criminal activity. The term for that is "spoofing."

Citizen Journalism – The Web and social tools provide the ability for any person with an Internet connection to publish information to the world. This concept is referred to as citizen journalism because any Internet Citizen can upload information without permission or filtering. We can literally "report" the news as we see it, or offer our editorial comments whenever we choose.

Cloud Computing – This is another broad term that has lots of different meanings depending on the context. For our purposes think of the cloud as a data center that allows people or organizations to rent or use computing power. There are a number of organizations that offer this service including Microsoft, IBM, Amazon and Google. Companies like Apple, Box.com and Dropbox offer data storage in the cloud for free, and for a fee when users need larger amounts of storage space.

Crowdsourcing – This term describes the process of using Websites to harness activities of a number of Internet users. There are crowdsourcing sites that help people raise money, create new logo's, develop new ideas, and solve problems. This is a powerful dynamic in that it is an early step towards tapping into the power of many people spread out all over the world. The concept of crowdsourcing is growing and is one of the early evidences of the Universal Mind.

Curate/Curation – Much like the meaning for curating in a museum, the technology version of this word refers to collecting, sorting, and archiving digital content. There are whole Websites dedicated to curation with an example being Pinterest. A Web user can curate digital music, pictures, video or any other type of content. Typically they will curate a narrow slice of digital content that they are interested in, for example, gathering only acoustic versions of digital songs.

Cyborg – When a human is augmented physically with technology we call that becoming a cyborg. There are varying degrees to this description because a person who has simply added one small piece of technology augmentation to their body can be described this way. Or, a person who has replaced 75% of their body can fall in the same category.

Database – The broad description of this term is a structured collection of data. We also use it to refer to specific company's tools that are used to store data, for example Microsoft SQL. This is a software application that is built specifically to store and manipulate large collections of data.

Decision Support System (DSS) – This describes a type of software that uses programmed rules to create guidance for making decisions. The rules that are programmed into the system are typically the same kinds of rules that person would go through when trying to make a decision. This allows the software to either make the decision outright, or be of assistance to a to a human decision maker.

Digital Content – When I use this term I am referring to the general spectrum of online videos, photo's, documents, presentations, etc.

Digital Courage – This term describes the increase of bravery in the way someone communicates when they are using online or digital tools. Because they do not have to face someone in person and deal with their possible reactions, the communicator feels safe delivering information they would normally not share.

Digital Marketing – This is another broad term that describes a wide range of marketing activities that involve tools like Websites, mobile applications, email marketing, and social tools. Digital Marketing uses data that is gathered on individual users in order to deliver content meant to influence behavior. The goal could be to sell a specific product, vote for a candidate, or take a specific action.

DIKW Chain – This is the name for the concept that data is improved to the state of information, and then knowledge and finally is turned into wisdom. We refer to it as a "chain" because as the data is transformed into the other states, the value increases.

Drones – This term is normally associated with pilotless flying objects. It actually can be used in a broader sense to describe any autonomous device that is remotely controlled by a human, or can be completely self controlled.

Dystopian/Utopian – These words are two sides of coin. Dystopian refers to a place where everything is destroyed, decrepit, and dark. Another variation if this term could be "post apocalyptic." Utopian describes a place where everything is as close to perfect as it can be.

Filtering/Filtering Systems – When I use these terms I am referring to software systems that are built specifically to separate streams of information into what is valuable to us versus what we would rather not see. Just as a filter in the physical world restricts some type of substance, a filter in the technology world restricts information you don't want to deal with.

Gesture Control – It is possible to control a computer or a device by using your hands to symbolize what you want done. The computer uses a camera to track the space where the user is located and follows gestures to replace using a mouse.

GPS (Global Positioning System) – This is the system controlled by satellites that provides the ability for a device to know where a specific location is on Earth. GPS is built into smartphones so that applications know where the user is physically and can provide directions to a location, or a listing of what is near the user.

Grid – When I use the phrases "on the grid or off the grid" I am referring to being connected to the Internet and all its tools, or being disconnected. Another way to think about this is whether a person is available to be contacted on their devices or not.

Hackers – There are actually different connotations for this word, some positive and others negative. In order to keep this simple, I have only used the word hackers to describe people who break into computer systems with ill intentions.

Haptic – This is a word that means that something is touch sensitive. It is normally associated with an interface that allows the control of a computing device by touching a screen, or simulating the touch of a screen.

Heads-up Display (HUD) – When a driver looks through a windshield and can see data projected on it, this is a heads-up display. When a gamer is moving through a gaming environment and can see data in the corners of the screen, that is another example. A HUD is simply data that is projected into your view as you are doing some other activity.

Hearable Device – This is a type of wearable device that is worn in the ear. It speaks to the user with information that has some context to where the user is geographically, or about what the user is doing at that moment. This allows for a flow of information to the user that allows them to be hands free and to not have anything distracting their vision.

Holographic – Holography is a three dimensional projection. Instead of a two dimensional projection on a screen, a holographic image is projected into an open space and closely simulates the physical subject being projected.

Humalogy – This word describes the combination of human and technology based activities to complete a process. The Humalogical Balance is a scale that allows us to rate how far on one side or the other a process, company, or system is towards being done by humans or technology. A simple example of this is the process of driving a car because half of this is accomplished by a machine and the other half is done through the activities of the driver.

Humanoid – This term can be used interchangeably with "cyborg." It generally refers to a human-looking technology based entity. In some cases it can refer to an alien being that looks somewhat like a human.

Implantable Technology – When technology is put inside the human body in some way it has been implanted. This can be be a Cochlear implant that helps deaf people hear, or could be sensors that are imbedded inside a persons hands in order to help them use gesture control with a device.

Infographic – This is a very popular way to deliver a collection of statistics or information. A static infographic is typically delivered as document file. Typically, the infographic is very colorful and based around pictures that are descriptive of the area of analytics being delivered. An active infographic is delivered as video with various statistics being scrolled across the screen.

Instagram – This is a popular social Website that is owned by Facebook. It provides a site for sharing photos with people in one's social network.

Interface(s) – The technology version of this term describes the the controlling elements that allow a user to work with software or devices. Whatever the method is for a user to work with a piece of technology is considered the interface between the user and the technology. This can be a screen for a piece of software, or it can be a keyboard or voice control for a smartphone.

Internet of Things – This term describes the new era of the Internet where devices become users of the Internet in order to communicate with each other or with users. As devices become "smart" and are able to deliver data to each other or their owners, they are creating an Internet of Things. It is likely that within a couple of years there will be more devices connected to the Internet than people.

Learning Ecosystem – This is an environment around a person that aids in their ability to learn. It can be carefully assembled so that it is robust and varied or it could be as simple as reading a book every once in a while. For the purposes of this book I am referring to a carefully crafted system for personal life-long learning.

Life-casting – There are people who document their life in great detail and then broadcast it to the public, or a small group. This can be done through an always-on video system or a device that takes pictures every minute or so of what the life-caster is seeing all day. The concept is to broadcast in a public way most of the details of one's life.

LinkedIn – This social site is used primarily as a business contact center. It provides the capability to connect to the 2nd degrees of separation people through using the network of 1st degree connection one builds.

Knowledgebase Systems – These software applications are built to house large amounts of information on specific topics. The are collections of all knowledge that can be extracted from people on a specific subject so that a user can do a search of the knowledgebase and retrieve answers to questions on that topic.

M2M (machine-to-machine) – This euphemism is another way to describe the Internet of Things. It refers to the capability for devices to talk to each other using an Internet connection.

Mashup (Data Mashup) – When two or more data sets are overlaid in order to see trends or anomalies there said to have been "mashed up." A popular example is mashing up mapping data with other forms of data in order to see geographic relevancy.

Nanobots – It will soon be possible to build rudimentary machines at nano scale (which is very tiny of course). The potential capabilities of doing this are staggering because nanobots would have the ability to repair or alter physical devices at a base level. For example, they could be used inside the human body to repair broken bones or damage to internal organs.

Neuroplasticity – This is the term we use to describe the fact that our brain can change shape to adapt to what we are using it for. When we call on our brain to do specific activities over and over again it will grow or shrink certain areas in order to better perform those tasks.

Oculus – This is the name of a company that has been developing improved virtual reality goggles. They were purchased by Facebook. This has lead many people to predict that Facebook will one day combine their social networking sites with the ability to act within a virtual space.

Operating System – Computers and devices must have a core piece of software that controls their capabilities. This software is called the operating system. Think of it like a translator that allows software applications to talk to the device or computer.

Outboard Brain – This is a euphemism for a mobile device. Because we use mobile devices to do activities that we have always done with our own brains (memorize information, solve problems, access information) the devices augment what our brain does. This is the reason the metaphor of an outboard brain fits.

Retinal Displays – Today we must look at a screen in order to interface with technology. A retinal display is a device that projects information from a device right into our vision. We have already built contact lenses that can display rudimentary information. The most valuable capability would be to push everything we see today on a computer screen right onto our retina.

Rules-based Systems – Software systems have the ability to have rules written into them so that they watch data flow and can take actions. In many ways our own brains work from a system of rules that we have learned over time. These rules dictate how we live our lives. It is possible to replicate some of these rules into software. For example, we know when we are in physical danger if a big person with a club yells at us and starts running in our direction. This is based on rules we have formed. Software can monitor financial transactions and can look for patterns that would tell it that a bad guy is trying to steal money. The software would be using rules to do the same.

Second Life – This is one of the most well known of the virtual world platforms. It is not a gaming environment; it is simply a virtual nation online. It has its own currency and land that is for sale. You are represented there by an avatar that you design when you sign up. It is a world with its own set of rules that sometimes mimic the real world and sometimes vary.

Self-learning Systems – It is now possible for software systems to observe user behavior and learn how to adjust actions. This can be as simply as noting when we make the same mistake over and over so the device learns to auto-correct it, or as complicated as "reading" huge amounts of documents so that the system can learn about a topic and make better decisions. The future potential of this is exciting and scary in that a self-learning system would have no boundaries as far as what it would learn unless we program them in.

Singularity – This word describes a concept developed by Ray Kurzweil that predicts that at some point this century, computers will grow in their capabilities to process information at the same rate as a human brain. Once that happens we would move into the Singularity where we have no idea what a collection of computers would think about and what they might be able to solve because they would far exceed our abilities to process information.

Smartphone/Devices – When a device has computer intelligence built into it so that it is much more capable than just being phone, it is called a smart phone. The other element that makes it smart is the ability to connect to the Internet.

Typically a smartphone has the ability to load software applications onto the device to extend its capabilities.

Snapchat – This social site has a unique capability to send messages between people that self destruct after a chosen amount of seconds. It is very popular with young people because it does not leave a history of the information provided to another person or group. For this reason, it is perfect for updating someone on what you are doing or thinking at a moment.

Social Listening Systems – These tools monitor the Web for any mention of keywords or phrases that the user chooses. The are often used by companies to track anything being said about them, their products or their employees. If someone on the Web posts anything that the user of this software is monitoring, the post will be recorded and the user is notified.

Social Technologies – This term is the descriptor for any Website or mobile application that provides a social networking or social media capability. This includes services like Facebook, Instagram, LinkedIn, and Snapchat.

Transhuman – This is the designation for the next stage of humanity after it has heavily integrated with technology. No one has definitively stated what the exact parameters are for having jumped from human to transhuman. That means it is a more of a theoretical concept that denotes that we will be very powerful when we integrate heavily with technology.

Universal Mind – This concept is also called the Hive Mind and it refers to a state of being where a large group of people are able to combine their minds in real time. The Universal Mind refers more to the ability for billions of people to connect minds in ways where we can both solve more complicated problems that could solved by individuals, communicate in real time with any other member, and flow love and support to members immediately when they need the help.

Viral Video – When a video has a high pass around rate between people so that it has a quickly escalating amount of views, it is described as viral. This refers to the compounding growth of views as people send links to all their friends, and then they forward to all their friends.

Virtuality – Let me reiterate, this is not an official word. This is something I made up to cover a broad category of technologies that provide some form of virtual display tool or application. They say only really creative people make up words. Note that we invented the word Humalogy as well!

Virtual Environment/Reality – This is the more traditional vocabulary for a virtual state. A virtual environment is a computer generated location. It could be a setting for a game, an office setting that hosts virtual meetings, or simply a virtual world like Second Life. Virtual reality is a general term for any 3D 360-degree view application.

Wearable Device – Any piece of technology that is constructed to be worn on the body is considered a wearable device. This includes smart watches and bands, smart clothing, and hearable devices. We will see an quickly escalating amount of variations of wearables over the next ten years.

Wetware – This is a slang term for the human body or some system that is tied into the human body.

CPSIA information can be obtained
at www.ICGtesting.com
Printed in the USA
LVOW01*0042070516

486732LV00007BA/18/P

9 780996 964975